Lecture Notes in Electrical Engineering

Volume 732

The book series *Lecture Notes in Electrical Engineering* (LNEE) publishes the latest developments in Electrical Engineering - quickly, informally and in high quality. While original research reported in proceedings and monographs has traditionally formed the core of LNEE, we also encourage authors to submit books devoted to supporting student education and professional training in the various fields and applications areas of electrical engineering. The series cover classical and emerging topics concerning:

- Communication Engineering, Information Theory and Networks
- Electronics Engineering and Microelectronics
- Signal, Image and Speech Processing
- Wireless and Mobile Communication
- Circuits and Systems
- Energy Systems, Power Electronics and Electrical Machines
- Electro-optical Engineering
- Instrumentation Engineering
- Avionics Engineering
- Control Systems
- Internet-of-Things and Cybersecurity
- Biomedical Devices, MEMS and NEMS

For general information about this book series, comments or suggestions, please contact leontina.dicecco@springer.com.

To submit a proposal or request further information, please contact the Publishing Editor in your country:

China

Jasmine Dou, Editor (jasmine.dou@springer.com)

India, Japan, Rest of Asia

Swati Meherishi, Editorial Director (Swati.Meherishi@springer.com)

Southeast Asia, Australia, New Zealand

Ramesh Nath Premnath, Editor (ramesh.premnath@springernature.com)

USA, Canada:

Michael Luby, Senior Editor (michael.luby@springer.com)

All other Countries:

Leontina Di Cecco, Senior Editor (leontina.dicecco@springer.com)

** This series is indexed by EI Compendex and Scopus databases.**

More information about this series at http://www.springer.com/series/7818

Karsten Berns · Alexander Köpper ·
Bernd Schürmann

Technical Foundations
of Embedded Systems

Electronics, System theory, Components
and Analysis

Springer

Karsten Berns
FB Informatik
TU Kaiserslautern
Kaiserslautern, Germany

Alexander Köpper
FB Informatik
TU Kaiserslautern
Kaiserslautern, Germany

Bernd Schürmann
FB Informatik
TU Kaiserslautern
Kaiserslautern, Germany

ISSN 1876-1100   ISSN 1876-1119 (electronic)
Lecture Notes in Electrical Engineering
ISBN 978-3-030-65159-6   ISBN 978-3-030-65157-2   (eBook)
https://doi.org/10.1007/978-3-030-65157-2

This Springer imprint is published by the registered company Springer Nature Switzerland AG
The registered company address is: Gewerbestrasse 11, 6330 Cham, Switzerland

Preface

Embedded systems are ubiquitous in the modern world. They can be found in smartphones, household appliances, modern motor vehicles or factories. They combine the analog with the digital world and perform a variety of different tasks. This means that developers of embedded systems must not only have extensive knowledge in designing special software systems, but also knowledge of electrical and control engineering, and have to deal with aspects of system components, communication and real-time requirements. This is the motivation for this textbook. The book summarizes the content of the lectures on Principles of Embedded Systems offered at the TU Kaiserslautern in the Department of Computer Science. It aims at students and users of computer science and at engineers, physicists and mathematicians, who are interested in the fundamentals of the development of embedded systems.

The book is divided into four parts to provide a basic structure for this. The first part deals with the principles of electrical engineering. Basic terms and methods are explained in such a way that students without extensive basic knowledge of electrical engineering have easy access to embedded systems. The second part discusses the fundamentals of systems theory and the design of controllers. The third part provides an overview of the system components of embedded systems and their integration. In the fourth part, modeling and analysis of the algorithms in relation to real-time requirements are explained. Some modeling techniques are introduced and their applications are described with examples, followed by methods of real-time planning.

This textbook was written with the cooperation of certain employees of the robotic system's chair, whom we would like to thank in particular at this point: Jens Hubrich for preparing the graphics and tables, Melanie Neudecker for taking the photographs and Axel Vierling and Christian Kötting for their technical additions. With this book, we hope to have made a significant contribution to the understanding of embedded systems. After reading the book, software developers will have a better understanding of the technical environment and its requirements. Conversely, engineers will have a deeper knowledge of the complexity of developing embedded software.

Kaiserslautern, Germany   Karsten Berns
April 2020   Alexander Köpper
                                                            Bernd Schürmann

Contents

Part IV   Modeling and Real-Time

Chapter 1
Introduction

Embedded systems are an integral part of the modern world. They can be found wherever direct interaction of technical systems with the environment takes place. They are used in various fields such as aerospace technology and consumer goods including televisions and washing machines. Hardly any technical device available today would be feasible without the use of embedded hardware and software systems. *Embedded systems* (*ES*) are information processing systems that are integrated into a larger product or environment (see Fig. 1.1).

This definition is deliberately rather vague, as the term "embedded systems" covers a wide range of applications. In building automation, for example, embedded systems are used for climate and light control, break-in protection and modern bell systems. Even in modern fitness trackers and smartphones, users interact with them constantly. Without digital, software-based processing of control and regulation problems and signal processing, today's driver assistance systems such as ABS, ESP, pedestrian and sign recognition and emergency brake assistance would be unthinkable. If such systems are networked in larger structures, they are also referred to as *Cyber Physical Systems* (*CPS*).

In general, embedded systems record internal states of the overall system as well as information about the environment via sensor systems. The measurement data obtained in this way are first processed by analog hardware and then processed by digital computer nodes. For these, in turn, special software has to be developed that must take both non-functional properties and functional requirements into account.

These include but are not limited to

- real-time requirements,
- weight, size, energy consumption,
- reliability, maintainability, availability,
- reliable real-time communication,
- connections for digital, analog and hybrid operating principles.

© The Author(s), under exclusive license to Springer Nature Switzerland AG 2021   1
K. Berns et al., *Technical Foundations of Embedded Systems*, Lecture Notes in Electrical
Engineering 732, https://doi.org/10.1007/978-3-030-65157-2_1

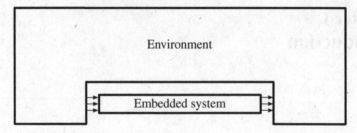

Fig. 1.1 Embedded system

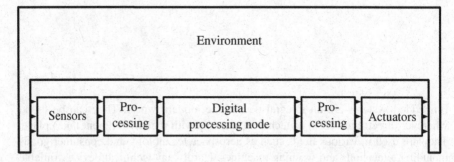

Fig. 1.2 Reactive embedded system

In contrast to software systems such as office applications, embedded systems usually process the programs cyclically (endless loop) and are optimized for specific applications. In the meantime, however, classic PC applications are shifting more to embedded systems, including from the point of view of software development, due in part to the use of complex operating systems and increasing resource availability.

Embedded systems can be built either as *reactive* (see Fig. 1.2) or as *transforming systems*. In reactive systems, the environment is changed by the control of actuators, e.g. the valve control on a boiler adjusts the temperature upwards or downwards. Transforming systems absorb data from the environment, process it digitally and then display it in modified form on an output device. An example of transforming systems is a modern TV set in which electromagnetic waves are first converted into electrical signals using a satellite dish and receiver, then digitally modeled and filtered and then output as video and audio data on the TV set.

Irrespective of their field of application, embedded systems are characterized above all by the intensive interaction between mechanics, electrical hardware and software. Only effective interaction between sensors, actuators and the software that is executed on microcontrollers or special hardware enables the complex functions available today to be delivered. The development of software-intensive embedded systems therefore requires both knowledge in the field of electrical engineering— such as the hardware used, systems theory and control engineering—and knowledge of systematic software development for complex systems from computer science. A good overview of the overlap between these adjacent fields is provided by e.g. [1, 2].

1.1   The Autonomous Forklift Truck as an Example System

Since an exclusively abstract consideration of the methods and components of embedded systems is often not sufficient to gain an understanding of how they work, practical examples are used in this book. In order to gain a better overview of the relationships, an example of a system will be presented below which is used to illustrate various aspects: the autonomous forklift truck.

This system is an autonomous vehicle with an electric engine on which a load fork that can be raised and lowered is mounted, similar to the one shown in Fig. 1.3. Various sensor systems such as distance sensors and cameras will be discussed in the course of the book. The system should be able to recognize and record objects in its environment autonomously and to move to specified destinations. As is does so, it should avoid obstacles in the surrounding area.

The system consists of sensors, actuators, processing units and the associated electrical and mechanical elements. Signals are transmitted in both analog and digital form. The system must satisfy non-functional requirements, such as avoiding obstacles that appear in good time, and must manage with a limited amount of energy (battery). A complex control system is required for the engine, as are communication interfaces between the components. These characteristics will be taken up and explored further in the following chapters.

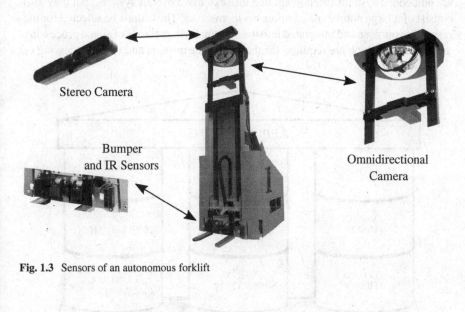

Stereo Camera

Bumper
and IR Sensors

Omnidirectional
Camera

Fig. 1.3   Sensors of an autonomous forklift

1.2 Structure of the Book

The electrical and electronic basics are explained in *part I*. Embedded systems consist of a variety of electrical and electronic components. From the sensor systems that convert physical input variables into digital data, through the actual processing unit in which these data are processed, to the electrically controlled actuators. In order to understand and calculate the processes taking place here and to identify possible sources of error, a sound basic knowledge of electrical engineering and electronics is required. Thus, they form the base of the structure in Fig. 1.4.

To this end, the principles of the calculation of electrical circuits (Chap. 2) are first introduced and then extended to electrical networks (Chap. 3). Then the functional principle of semiconductor devices is introduced (Chap. 4) and their practical application is explained, including their usage in operational amplifier circuits.

In *part II*, systems theory and control engineering are examined. Control systems are required to react correctly to unpredictable inputs, especially for time-critical applications. First, the formal and mathematical basics used for this are developed within the framework of *systems theory* (Chap. 5). Building on this, the concrete procedure for the design of classical and fuzzy controllers is then presented (Chap. 6). These chapters form most of the leftmost column in Fig. 1.4.

In *part III*, system components of embedded systems are presented. Not only are embedded systems often integrated into extensive overall systems, but they also consist of a large number of components themselves. These must be selected for the respective purpose and integrated into the system. First, methods of signal processing are explained which are required for the basic transmission and (pre-)processing of

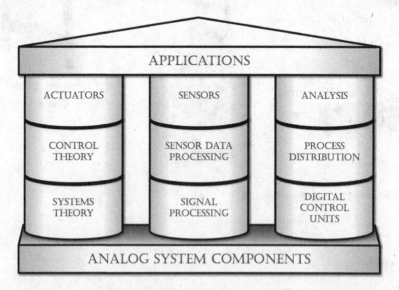

Fig. 1.4 The structure of the book including the dependencies of the discussed topics

incoming and outgoing signals (Chap. 7). Then the term sensor is explained and specific processing of sensor data is dealt with (Chap. 8). Various sensor types are presented for this purpose. These chapters can be found in the middle column of Fig. 1.4. Certain actuators (Chap. 9) and various architectures of the digital control unit (Chap. 10) are also introduced. Finally, abstract communication of embedded systems with the environment and in the system itself is discussed (Chap. 11).

Part IV deals with modeling of real-time behavior. Embedded systems often have to meet certain quality criteria. In addition to a failsafe reliability and the actual function, it is usually necessary to meet efficiency and time specifications. Extensive knowledge of the processes in hardware and software is required to achieve this. Abstract modeling and analysis methods are introduced (Chap. 12) as a way of acquiring this. On this basis, different methods are presented for meeting real-time requirements (Chap. 13). These topics are shown in the right column of Fig. 1.4.

References

1. Marwedel P (2011) Embedded system design, 2nd edn. Springer, Dordrecht [i.a.]
2. Bräunl T (2003) Embedded robotics: mobile robot design and applications with embedded systems. Springer, Berlin [i.a.]

Part I
Electrical and Electronic Basics

Chapter 2
Electrotechnical Basics

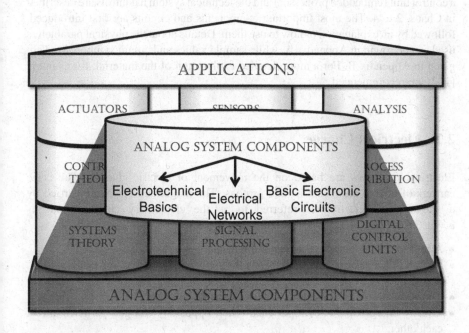

This chapter introduces the most important fundamental terms and the basic physical principles of electrical engineering. The use of important parameters, such as energy and power in the electrical circuit, is examined, as is resistance as a central component of electrical circuits. This establishes the prerequisites for calculations for electrical circuits, which are used, for example, in signal processing. Finally, ohmic resistance is presented as the most important element of electrical circuits. Among other things, this is used for the analysis of electrical networks, which the next chapter deals with.

© The Author(s), under exclusive license to Springer Nature Switzerland AG 2021
K. Berns et al., *Technical Foundations of Embedded Systems*, Lecture Notes in Electrical
Engineering 732, https://doi.org/10.1007/978-3-030-65157-2_2

2.1 Electrotechnical Consideration of Embedded Systems

Software developers working in the field of embedded systems need in-depth knowledge of the technical system for which they are designing their embedded software. As resources are often limited, they need an understanding of how these resources work to make the most of them. Furthermore, many embedded systems do not have certain abstraction layers available, such as the operating system or a driver layer, making it necessary to work very closely with the hardware. It is therefore essential to have a basic understanding of the hardware and the concepts used. This also serves as a basis for mutual understanding within the usually interdisciplinary development team for such systems, since communication in practice often constitutes a bottleneck in system development.

The basic electrical and electronic interrelationships necessary for subsequent understanding of control engineering, sensor processing and the interfaces between a control unit (embedded processor) and the technical system it controls are described in Chaps. 2 to 4. The most important components and circuits are first introduced, followed by an explanation of how to use them. Details about the physical parameters used can be found in Appendix A, while sample values and circuit symbols used are given in Appendix B. For a more detailed explanation of the material, for example [1, 2] are recommended.

2.2 Electrical Charge

Electrical processes are based on the movement of electric charges. The basic carriers of charge are electrons and protons. The magnitude of charge carried by a single proton/electron is referred to as the elementary charge e and is $e = 1.602176634 \cdot 10^{-19}$ As. The following applies:

- A proton has the charge e, an electron the charge $-e$.
- The unit of measurement of charge is the *Ampere second* [As] or *Coulomb* [C], where 1 As = 1 C.
- The abbreviation Q is used for electrical charge.
- Equally charged particles repel each other, oppositely charged particles attract each other.

The electrical charge, here as an accumulation of many charge carriers, plays an important role in using components that can store electrical energy in particular, such as batteries and capacitors. All electrically charged bodies are surrounded by an *electric field*. This acts as a force field on charges and thus influences their behavior.

2.3 Electric Current

Electric current is the directional movement of charges. These can be electrons or ions. The charges are transported via (semi-)conductors, which are substances with many mobile charge carriers.

The *electric current strength I*, also referred to as the *amperage* or simply *current*, is the amount of charge Q, that flows through a fixed cross-section of a conductor per unit of time t. The unit of current is the *Ampere* [A], named after André-Marie Ampère, who established the theory of electromagnetism around 1820. Current strength is a directional parameter, so the direction of flow of the charge carriers influences the calculations. The calculation is always based on the technical current direction, which by definition is the direction of flow of the (imaginary) positive charge carriers. Negatively charged electrons therefore flow in the opposite direction to the technical current direction. Although in physical reality it is the negatively charged electrons that usually flow (exceptions would be e.g. positively charged ions in current-carrying liquids), this convention has prevailed historically.

A temporally constant current equals a constant charge flow over time, yielding

$$I = \frac{Q}{t} \text{ resp. } Q = I \cdot t \tag{2.1}$$

If the current $I(t)$ is not constant over the time interval $[t_1, t_2]$ under consideration, the transported charge must be calculated section by section. For a temporally variable current, the approximation of N time intervals gives the duration Δt, in which $I(t)$ is constant

$$Q_i = I(t_i) \cdot \Delta t$$

The transition from $N \to \infty$ or $\Delta t \to 0$ leads to the integral

$$Q_{1,2} = \int_{t_1}^{t_2} I(t)dt$$

for charge carrier transport in the time interval t_1 to t_2. With an initial charge Q_0, the charge that has flowed at the time τ is

$$Q(\tau) = \int_{t=0}^{\tau} I(t)dt + Q_0 \tag{2.2}$$

Conversely, the current strength results from the charge transport as a function of time

$$I(t) = \frac{d}{dt} Q(t) \tag{2.3}$$

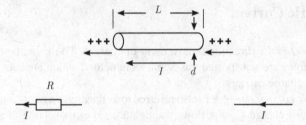

Fig. 2.1 Graphical representations of electric current in a wire

The representation of a current through a conductor is usually simplified as an arrow on the conductor itself, as shown in Fig. 2.1.

2.4   Electrical Resistance

When electrons move in a conductor, they repeatedly collide with its atoms, as a result of which their energy is released in the form of heat. Electrical resistance R describes how easy or difficult it is for electrons to move in a conductor. The unit of resistance is the *Ohm* [Ω] where $1\ \Omega = 1\,V/A$. The electrical component that implements an electrical resistance as a circuit element is called *resistor*.

The electrical resistance increases in proportion to the length l and in inverse proportion to the cross-section q of the conductor. In addition, the resistance depends on the temperature and the material of the conductor, which is described by the specific resistance ρ.

$$R = \rho \cdot \frac{l}{q} \tag{2.4}$$

A list of different materials and their specific resistance can be found in the Appendix in Table B.1. Since (ohmic) resistances occur in every electrical component, they are an essential element of electrical engineering. Their circuit symbol is shown in Fig. 2.2.

2.5   The Electrical Circuit

Current flows only in a closed circuit (see Fig. 2.3). A source (voltage or current source) always generates the same number of positive and negative charges, with the

Fig. 2.2 Graphical symbol of the electrical resistance/resistor

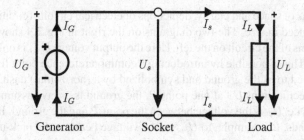

Fig. 2.3 Simple electrical circuit with generator voltage U_G, voltage U_s and load voltage U_L

moving charge carriers flowing through the circuit. If the circuit is not branched, the current strength is the same at each point.

The *electrical voltage* (U) is the quotient of the work required to shift a charge Q from a to b and the charge Q itself.

$$U = \frac{W_{ab}}{Q} \tag{2.5}$$

The unit of voltage is the *Volt* [V] where $1\,V = 1^W/_A = 1^{Nm}/_{As}$, named after Alessandro Volta, who, among other things, designed the Voltaic pile, a precursor of today's batteries, in 1780. The following applies to the voltage:

- A voltage is always defined between two points only.
- A voltage always has a direction, indicated on the circuit diagram by the voltage arrow.
- Current is represented by an arrow on the conductor, voltage by an arrow next to the conductor/component or between the terminals.

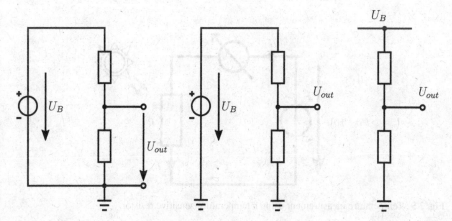

Fig. 2.4 Possible simplifications of the circuit diagram in Fig. 2.3

For reasons of space and clarity, depictions of electrical circuits (circuit diagrams) are often reduced in size. The two diagrams on the right in Fig. 2.4 show simplified representations of the circuit on the left. Here the output voltage U_{out} is only displayed at one point. This is possible by introducing a common reference point for the whole circuit, referred to as the ground and symbolized by a (horizontal) dash. In the case of voltage specifications U_P at one point P, the ground is always assumed to be the second point, i.e. U_P is the voltage between the point P and the ground. In the figure, this corresponds, for example, to U_{out} as the voltage between the pick-up point and the ground. If several ground points are marked on a circuit, they must be connected to each other or assumed to be connected.

2.6 Ohm's Law

The relationship between current I and voltage U at a resistor R is described by *Ohm's law*, named after the physicist Georg Simon Ohm, who demonstrated it in 1826.

$$U = R \cdot I \tag{2.6}$$

This law also applies to time-dependent variables (e.g. AC voltage, variable resistance), but R is only constant with ohmic resistances. In reality, resistance may be voltage-dependent and temperature-dependent. Ohm's law is the basic law for the calculation of electrical networks.

With the knowledge acquired thus far, a simple temperature measurement can be carried out using a current meter. The layout for this is shown in Fig. 2.5. The resistance R at a temperature of 20 °C is 100Ω and at 40 °C 110Ω. The voltage source supplies 10 V. This would cause the current meter to display $I = \frac{U}{R} = \frac{10V}{100\,\Omega} = 0.1$ A at 20 °C and 0.09 A at 40 °C. The measured current can thus be converted easily to the temperature of the resistor.

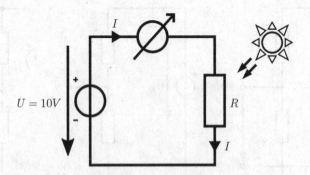

Fig. 2.5 Temperature measurement using a temperature-sensitive resistor

2.7 Energy and Power

According to the law of conservation of energy, the total energy in a closed system always remains the same. No energy is lost and no energy can be created. Energy is only converted between different forms, e.g. from electrical to mechanical energy and vice versa. This process plays an important role in embedded systems, since such processes take place both in transduction (see Sect. 7.2) and in the actuators (see Chap. 9). In addition to these intended effects, the unwanted conversion of electrical energy into thermal energy due to the electrical resistance of the components must often also be taken into account (see Sect. 2.4). On the one hand, this leads to higher energy consumption, which can be particularly relevant for battery-powered systems. On the other hand, excessive heating can destroy components, so that a cooling mechanism may have to be included.

The voltage between two points P_1 and P_2 corresponds to the work or energy, that must be applied to move a charge Q from P_1 to P_2.

$$W = U \cdot Q \tag{2.7}$$

$$W = \int_{t=0}^{T} U(t) \cdot I(t)dt \tag{2.8}$$

or $W = U \cdot I \cdot t$ at a constant voltage and current. The unit of energy or work W is the *Joule* [J], where $1\,J = 1\,Ws = 1\,VAs$, irrespective of the form of energy. However, electrical energy is usually given in [Ws] and not in the general energy unit [J], which is usual for thermal energy.

All forms of energy (electrical energy, thermal energy, kinetic energy,...) are equivalent. For example, an electric motor converts electrical energy into kinetic energy, a brake converts kinetic energy into thermal energy and a battery converts chemical energy into electrical energy.

$$W_{electrical} \equiv W_{kinetic} \equiv W_{thermal} \equiv W_{potential} \equiv W_{...}$$

When comparing different systems in terms of their performance and consumption, energy is only suitable to a limited extent due to its dependency on time. Here the consideration of *power* helps. The (electrical) power is the (electrical) work applied per unit of time t.

$$P(t) = \frac{dW(t)}{dt} \Rightarrow P = U \cdot I \text{ (for direct current/voltage)} \tag{2.9}$$

The electrical power is given in *Watt* [W], where $1\,W = 1\,VA$. Ohm's law can be used to calculate the power of a resistor to

$$P = U \cdot I = \frac{U^2}{R} = I^2 \cdot R \qquad (2.10)$$

This formula is used for a realistic estimate of the consumption of components. Information about the voltage or the current which falls/flows through a component alone is not sufficient without knowing the other variable in this connection. These are already combined in the power as instantaneous values.

2.8   Time-Variable Voltages

In the previous sections, the focus was essentially on currents and voltages that are constant over time. However, most electrical signals in embedded systems are variable over time. This is important in the area of sensor technology in particular (see e.g. Sect. 8.5.2). A distinction is made between three types of current or voltage: Direct current/voltage (time constant), usually referred to as DC; alternating current/voltage (time variable, mean value 0), often referred to as AC and variable/mixed current/voltage (time variable, mean value not equal to 0). Pure direct currents/voltages are described completely by a single value. Periodic alternating currents/voltages can be described by superimposing sinusoidal functions with different frequencies. If the DC voltage predominates in a mixed voltage, this is also referred to as a DC voltage with an AC voltage component or ripple (see Fig. 2.6).

To describe AC voltages, the frequency of the voltage is required in addition to the amplitude or the effective value (see below). The *frequency* f is the reciprocal of the period T with the unit *Hertz* [Hz] and a conversion of 1 Hz = $1^1/s$. If

$$U(t + T) = U(t) \; \forall t \qquad (2.11)$$

is a periodic voltage, the respective frequency f results to

$$f = \frac{1}{T} \qquad (2.12)$$

An example is a socket with a sinusoidal AC voltage which changes direction 100 times per second. A period lasts 20 ms, so the frequency is 50 Hz. The amplitude of the voltage is 325 V. The change over time is shown in Fig. 2.7.

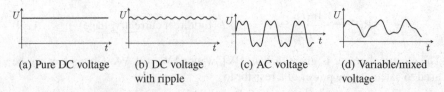

(a) Pure DC voltage   (b) DC voltage   (c) AC voltage   (d) Variable/mixed
                          with ripple   voltage

Fig. 2.6 Courses of voltages

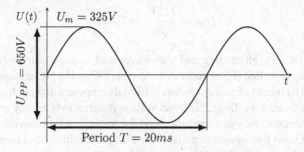

Fig. 2.7 Temporal course of an AC voltage: U_m depicts the amplitude, U_{PP} the peak-to-peak voltage

The function of the mains AC voltage is

$$U(t) = U_m \cdot \sin(2\pi \cdot f \cdot t)$$

The AC voltage at the socket is always specified by the *effective voltage* 230 V. This is the voltage that a DC voltage would have to have to make a light bulb shine as brightly as the given AC voltage does. With sinusoidal AC voltages, the effective voltage U_{eff} and amplitude U_m are related according to the following equation

$$U_{\text{eff}} = \frac{U_m}{\sqrt{2}} \approx 0.707 \cdot U_m \tag{2.13}$$

For the European synchronous grid, this yields

Table 2.1 Examples of time dependent voltage curves

Temporal course	Use-case
	Determination of a networks' behavior during switching processes
	To display switching on and off processes or as clock generator for digital circuits
	To measure the linearity of networks or to record nonlinear characteristic curves
	Sine function with adjustable frequency to measure the frequency response of networks
	Addition to test signals with superimposed noise. Often with a very long period

$$U_{\text{eff}} = \frac{325\,V}{\sqrt{2}} \approx 230V$$

In addition to direct, alternating and variable/pulsed voltages, any signals can be generated using so-called signal generators. Several DC and/or AC voltage signals are superimposed by means of analog circuits or digital computers in order to generate the new signal (cf. Sect. 5.4). Table 2.1 shows voltage progressions as they are generated by signal generators; they are typically used for testing and measuring circuits. In particular, the *jump function* shown first plays a major role in control technology and will be referred to frequently in the course of the book.

References

1. Svoboda JA, Richard CD (2013) Introduction to electric circuits. Wiley, Jefferson City (2013)
2. Chen WK (ed) (2004) The electrical engineering handbook. Elsevier, Burlington

Chapter 3
Electrical Networks

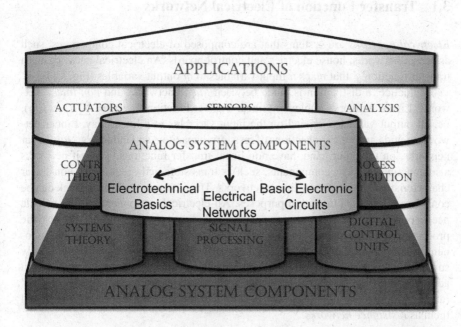

This chapter introduces the calculation of electrical networks. The most important circuit rules make it possible to determine the voltages and currents at different locations in the network by means of abstraction to resistance networks and two or four poles. Voltage and current dividers allow different voltages and currents to be generated. This is an important prerequisite for the construction of measuring bridges and amplifier circuits, for example (see Sects. 8.5.2 and 4.5.1).

Furthermore, the capacitors and coils are introduced as components and their behavior in the electrical network is examined more closely. These components can be used to store analog states. This can be used to build frequency-dependent and oscillating circuits (see Sect. 6.2.10) and to model the time dependence of certain networks. As an example, the low pass is introduced as an important filter circuit.

© The Author(s), under exclusive license to Springer Nature Switzerland AG 2021
K. Berns et al., *Technical Foundations of Embedded Systems*, Lecture Notes in Electrical
Engineering 732, https://doi.org/10.1007/978-3-030-65157-2_3

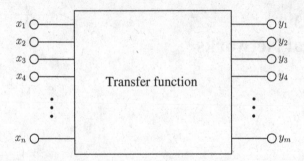

Fig. 3.1 Generalized electrical network

3.1   Transfer Function of Electrical Networks

Electrical networks are systems that are composed of electrical components, such as sensor networks, house electrics and control boards. An electrical network has a transfer function f that maps n input variables to m output variables (Fig. 3.1).

In practice, a distinction is made between *linear* networks and *non-linear* networks. Linear networks implement linear transfer functions ($y_i(t) = \sum a_j \cdot x_j(t)$), i.e. all output variables depend on the input variables in a linear way. Linear networks include any interconnections of resistors, capacitors and inductors. Non-linear networks, on the other hand, have non-linear transfer functions. Thus, if networks also contain nonlinear components, such as transistors and diodes with non-linear characteristics, these are non-linear networks. The autonomous forklift truck can be considered here for illustration purposes. All electrical components of the vehicle need a power supply. The power consumers include not only the motors, but also the processing unit, the sensors and all the displays and lights. However, these components do not all require the same supply voltage. While processing units are usually supplied with 1.8–5 V, several hundred volts are not unusual for electric motors. In addition, there are the signal transmission levels, which tend to be in the 5–12 V range. In order to be able to analyze networks of this sort, one approach is to model them as *resistance networks*.

3.2   Resistance Networks

Resistance networks are circuits of any complexity consisting exclusively of resistors, current sources and voltage sources. Analysis of such networks, i.e. the calculation of all currents and voltages, takes place on the basis of *Kirchhoff's circuit laws*, the nodal rule and the mesh rule, which Gustav Robert Kirchhoff formulated in 1845. These networks and rules are of major importance for electrical engineering. By transforming any electrical networks into resistance networks, it is possible to analyze even complex circuits (see below) with comparatively little effort, at least for certain basic conditions.

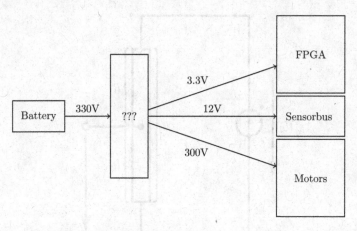

Fig. 3.2 Example: voltage supply of the forklift truck

The following scenario depicts as an example for the autonomous forklift truck (see Fig. 3.2). An FPGA (see Sect. 10.5) is used as a processing node to drive a motor. It is connected to the motor output stage by means of a bus system. A PC is also connected here for external control. However, the energy supply is problematic. The FPGA requires a supply voltage of 3.3 V or 1.8 V; the motor itself is driven by 300 V and both the bus system and the PC interface assume a 12 V communication level. Supply of the required voltages from outside would lead to many cables and thus a heavy weight, on the one hand, and on the other hand would considerably restrict the mobility of the forklift truck. Thus, all supply voltages have to be taken from an on-board 330 V battery.

Below, a voltage divider is used to provide the option of generating the different voltages required from the battery voltage. Although this is in practice not the preferred way to solve this problem, it offers a good overview of the idea.

3.2.1 Series Connection of Resistors

A series connection of R_1 and R_2 can be regarded as *tapping* into a total resistance R (see Fig. 3.3). R has a length l, R_1 and R_2 have the lengths l_1 and l_2 respectively, with $l = l_1 + l_2$.

Then according to formula 2.4, it follows

$$R = \rho \cdot \frac{l}{q} = \frac{\rho}{q}(l_1 + l_2)$$

$$R_1 = \frac{\rho \cdot l_1}{q} \text{ and } R_2 = \frac{\rho \cdot l_2}{q}$$

$$\Rightarrow R = R_1 + R_2 \tag{3.1}$$

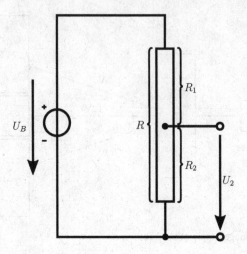

Fig. 3.3 Tapping of a resistor

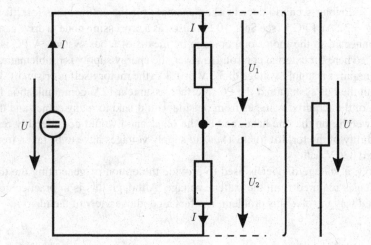

Fig. 3.4 Series connection of resistors

This leads to

$$U = I \cdot R = I(R_1 + R_2)$$
$$U_1 = I \cdot R_1 \text{ and } U_2 = I \cdot R_2$$
$$\Rightarrow U = U_1 + U_2 \tag{3.2}$$

In a series connection of resistors, the resistors and the voltages across the resistors are added together, see Fig. 3.4. The voltage U_2 that can be obtained with a *voltage divider* (Potentiometer) from U is often of interest. From

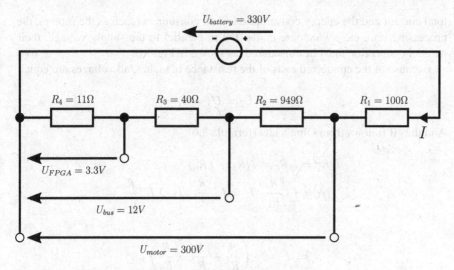

Fig. 3.5 Example: voltage divider for the supply of the forklift truck

$$I = \frac{U}{R_1 + R_2} \text{ and } U_2 = I \cdot R_2 \tag{3.3}$$

the *voltage divider rule* is derived

$$U_2 = U \cdot \frac{R_2}{R_1 + R_2} \tag{3.4}$$

If this consideration is extended to series circuits consisting of several resistors, the following general formula can be derived (see also Fig. 3.4).

$$U_2 = U \cdot \frac{R_x}{R_{tot}}, \text{ with } R_x = \sum_i^N R_i \tag{3.5}$$

This also solves the problem of different supply voltages. The largest occurring voltage is selected as the supply voltage U. All smaller voltages can now be supplied according to formula 3.5 by means of taps between resistors of appropriate size from a voltage divider (see Fig. 3.5).

3.2.2 Parallel Connection of Resistors

In line with the series connection of resistors introduced above, calculation rules also exist for their connection in parallel. Such parallel circuits emerge, for example, if several loads are connected to the same power supply. In order to calculate the total current and the energy consumption of all consumers (such as the motors, the processing unit, etc.) which are connected in parallel to the supply voltage, their parallel connection must be considered, as shown in Fig. 3.6.

Because of the connected ends of the resistance branches, all voltages are equal.

$$U_1 = U_2 = U_3 = U$$

And thus it follows from Ohm's law (formula 2.6)

$$I_1 R_1 = I_2 R_2 = I_3 R_3 = I R_{tot}$$

$$I_1 = I \cdot \frac{R}{R_1} \quad I_2 = I \cdot \frac{R}{R_2} \quad I_3 = I \cdot \frac{R}{R_3}$$

$$\Rightarrow I = I \cdot \left(\frac{R}{R_1} + \frac{R}{R_2} + \frac{R}{R_3} \right)$$

$$\Rightarrow \frac{1}{R} = \frac{1}{R_1} + \frac{1}{R_2} + \frac{1}{R_3}$$

The general equation for N resistors connected in parallel is

$$\frac{1}{R} = \sum_{i=1}^{N} \frac{1}{R_i} \tag{3.6}$$

Here, $\frac{1}{R}$ is called the *conductance* of R. The unit of conductance is the *Siemens* [S], where $1\,S = 1\,1/\Omega$, named after Werner von Siemens, who developed the first electrical generator in 1866 and is considered the founder of electrical power engineering.

When resistors are connected in parallel, the reciprocal values of the resistors add up to the reciprocal value of the total resistance. So for two resistors connected in

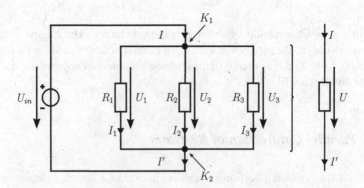

Fig. 3.6 Parallel connection of resistors

parallel, the overall resistance becomes

$$R = R_1 || R_2 = \frac{R_1 \cdot R_2}{R_1 + R_2} \quad , \text{with } || \text{ meaning } parallel\,to.$$

$$\frac{I_1}{I_2} = \frac{R_2}{R_1} \quad (current\,divider\,rule) \tag{3.7}$$

This formula plays an important role in network analysis, especially in energy consumption estimations. With more complex networks, however, the calculations quickly become unclear with only voltage divider and current divider rules. A simplification can be achieved by using the two Kirchhoff laws, which are valid not only for linear networks.

3.2.3 Kirchhoff's First Law (Nodal Rule)

Parallel circuits consist of several branches, so-called nodes. All branches between which there are no components can be combined to form a node. Since the currents branch at these nodes, but no charge is added or discharged from outside the connected branches, the following applies.

In each node of a circuit, the sum of the incoming currents is equal to the sum of the outgoing currents (see Fig. 3.7). Since the direction of the current arrow must be observed, this corresponds to the following statement.

The sum of all currents that flow into a node is zero (*nodal rule*, also called *current law*).

$$\sum_{i=1}^{N} I_i = 0 \tag{3.8}$$

The parallel connection of resistors in Fig. 3.6 is a good example. It results in the corresponding node equations for the network.

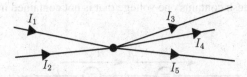

Fig. 3.7 It results $I_1 + I_2 = I_3 + I_4 + I_5$ or $I_1 + I_2 - I_3 - I_4 - I_5 = 0$

$$K_1 : I = I_1 + I_2 + I_3$$
$$K_2 : I_1 + I_2 + I_3 = I'$$

The second node equation is superfluous in this case, since $I = I'$.

3.2.4   Kirchhoff's Second Law (Mesh Rule)

A closed circuit in a network is called a mesh. Such a mesh can be any loop through the network, where the endpoint coincides with the beginning. In such a mesh, the sum of all source voltages must be equal to the sum of all voltage drops. Or more generally:

In each mesh of a network, the sum of all voltages is zero (*mesh rule*).

$$\sum_{i=1}^{N} U_i = 0 \quad \text{for } N \text{ voltages in a mesh} \tag{3.9}$$

As an example, the mesh rule can be applied to a simple resistance network (Fig. 3.8).

The direction of the *mesh arrow* for the mesh is determined arbitrarily. All voltages in the direction of the mesh arrow are calculated positively, all others negatively. If there are several meshes, all mesh arrows should have the *same direction of rotation*. Otherwise, the signs of the equations obtained would have to be adjusted repeatedly, which could easily lead to calculation errors. The relation between the voltages in the mesh is thus

$$U_{s1} + U_{s2} - U_1 - U_2 - U_3 = 0 \tag{3.10}$$

This equation alone is not sufficient to analyze the network. To analyze a resistance network, the (independent) node and mesh equations (explained below) are combined into a system of equations and then solved.

1. First, all independent meshes are determined. Each voltage must be contained in (at least) one mesh. It is recommended on the other hand to chose the meshes in a way that each mesh contains one voltage that is not contained in any other mesh.

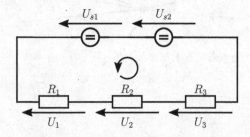

Fig. 3.8 Simple network with mesh arrows (middle of the picture)

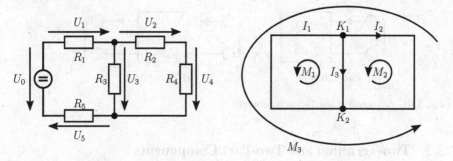

Fig. 3.9 Network with three meshes and mesh arrows

2. Then node equations are added until the number of equations equals the number of unknown currents and voltages.

This approach to determining currents and voltages in a network can also be applied to more general networks with non-linear devices, such as diodes and transistors.

A three-mesh resistor network (Fig. 3.9) provides an example. U_0, R_1, R_2, R_3, R_4, R_5 are given, the branch currents I_1, I_2, I_3 are to be determined.

First the meshes M_1, M_2 and M_3 are selected as shown in the right image in Fig. 3.9. The following mesh equations result.

$$M_1 : U_0 - U_1 - U_3 - U_5 = 0$$
$$M_2 : U_3 - U_2 - U_4 = 0$$
$$M_3 : U_0 - U_1 - U_2 - U_4 - U_5 = 0$$

By replacing the voltages across the resistors according to Ohm's law (formula 2.6) we obtain

$$M_1 : U_0 - I_1 R_1 - I_3 R_3 - I_1 R_5 = 0$$
$$M_2 : I_3 R_3 - I_2 R_2 - I_2 R_4 = 0$$
$$M_3 : U_0 - I_1 R_1 - I_2 R_2 - I_2 R_4 - I_1 R_5 = 0$$

Subtraction $M_3 - M_1$ results in

$$I_3 R_3 - I_2 R_2 - I_2 R_4 = 0 \quad \text{(equals } M_2)$$

This means that the system of equations is not linearly independent, since no new voltages were recorded with M_3. A third, independent equation is therefore required to calculate the currents.

$$K_1 : I_1 - I_2 - I_3 = 0 \quad \text{(nodal rule)}$$

The solution is now obtained by Gaussian elimination applied to a linearly independent system of equations (e.g. M_1, M_2, K_1).

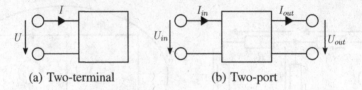

(a) Two-terminal   (b) Two-port

Fig. 3.10 Two-terminal and Two-port networks

3.3   Two-Terminal and Two-Port Components

Frequently used special cases of electrical networks are two-terminal and two-port networks, i.e. networks with two connections or one input and one output pair. These often serve as a simplified abstraction of networks at a lower level of abstraction.

In Fig. 3.10 a two-terminal and a two-port network are shown. A two-terminal network (e.g. a resistor R) can be described entirely by the relation between current I and voltage U. A two-port network(special case of a quadripole/four-terminal network) is uniquely described by U_{in}, U_{out}, I_{in}, I_{out} and the four relations between these values.

Two- or four-terminal networks are called *active*, if they contain an energy source (careful: These are often not explicitly drawn in abstract diagrams, see Sect. 4.7.3), otherwise they are called *passive*.

3.4   Resistor-Capacitor Networks

In electrical engineering, two other passive components are important in addition to the resistor previously considered: the capacitor (capacitance) and the coil/inductor (inductance/inductivity). These are mainly used to store energy and filter certain frequencies (see Sect. 3.5) in alternating signals. They are used in these areas in almost all embedded systems. Here only the capacitors in the time range are described, as representatives for both components. The coil behaves as a mirror image for most phenomena described in Sect. 3.6.

In principle, a capacitor consists of two metal plates (electrodes) which are separated by an insulator, also known as a dielectric (see Fig. 3.11). The metal plates can store electrical charges, i.e. electrical energy. The capacity C of a capacitor indicates how many charge carriers the capacitor can accommodate. This depends on the surface area A and the distance d between the metal plates and the insulation material. Since the charge carriers accumulate on the metal surface, the capacity is independent of the thickness of the metal plates. Capacitors are used in circuits, e.g. to smooth out voltages or to reduce interference pulses. Capacitances also occur unintentionally wherever electrical conductors run close together (e.g. on ICs), or where different voltage levels occur.

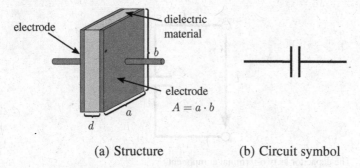

(a) Structure   (b) Circuit symbol

Fig. 3.11 Structure and graphical symbol of the capacitor

Capacities have a strong influence on the time response of circuits. The unit of capacity is the *Farad* [F], where 1 F = 1 As/V, named after Michael Faraday, who discovered electromagnetic induction in 1831, among other things. The capacitance of a capacitor is calculated as

$$C = \varepsilon_0 \cdot \varepsilon_r \cdot \frac{A}{d} \tag{3.11}$$

The relative dielectric constant ε_r describes the insulation material. In a vacuum it is $\varepsilon_r = 1$, with air it is $\varepsilon_r = 1.006$ and with SiO_2, a commonly used insulation material in microelectronics, it is $\varepsilon_r = 3.9$. ε_0 is the proportionality factor that adapts the capacitance equation to the SI system of units ($\varepsilon_0 = 8.8 \cdot 10^{-12}$ F/m). The capacities used in electronics are usually very small. They lie in the order of magnitude of the picofarad (pF) to the millifarad (mF, for the prefix cf. Appendix, Table A.2).

3.4.1 The Capacitor as a Two-Terminal Component

In order to apply a charge Q to a capacitor, a voltage is required which is applied to the capacitor. This voltage causes charge carriers to flow onto the capacitor. These form an electric field in the capacitor, which counteracts the applied voltage and over time ends the charging process. The voltage required to bring further positive charge carriers to the positive side increases with the number of charge carriers already present (repulsion of identical charges). In Fig. 3.12 it is assumed that the voltage U between the open terminals is fixed. This could be caused, for example, by a connected voltage source or a tap in a voltage divider. The current I transports charge carriers to the capacitor with the capacity C. The greater the capacitance of a capacitor or the voltage applied, the greater the charge on the capacitor.

$$Q = C \cdot U \tag{3.12}$$

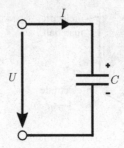

Fig. 3.12 The capacitor as two-terminal component

3.4.1.1   Voltage and Current Characteristics of Capacitors

This relationship is particularly interesting with regard to the current curve of a capacitor. The following results from formula 2.2.

$$Q(t) = \int_0^t I(t)dt + Q_0$$

$$I(t) = \frac{d}{dt}Q(t)$$

with $Q = C \cdot U$ it follows

$$U(t) = \frac{1}{C} \int_0^t I(t)dt + \frac{Q_0}{C} \tag{3.13}$$

$$I(t) = C \cdot \frac{dU(t)}{dt} \tag{3.14}$$

This function of the temporal dependence of current and voltage on a capacitor allows calculation of circuits that contain these components, such as the low pass (see Sect. 3.5.1) and the integrator (see Sect. 4.10.3).

3.4.2   Parallel Connection of Capacitors

Like resistors, capacitors can also be connected in parallel and in series. The total capacitance C of the capacitors C_1, C_2 connected in parallel is now determined in Fig. 3.13. From Eq. 3.12 it follows

$$Q_1 = C_1 \cdot U \text{ and } Q_2 = C_2 \cdot U$$

Therefore, the total charge results to

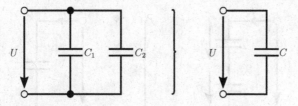

Fig. 3.13 Parallel connection of capacitors and equivalent circuit diagram

$$Q = Q_1 + Q_2$$
$$= C_1 U + C_2 U$$
$$= (C_1 + C_2)U$$
$$= CU$$
$$\Rightarrow C = C_1 + C_2 \tag{3.15}$$

When capacitors are connected in parallel, the surface area increases. The total capacity is the sum of the individual capacities.

$$C_{tot} = \sum_{i=1}^{N} C_i \tag{3.16}$$

3.4.3 Series Connection of Capacitors

Figure 3.14 shows the series connection of capacitors. Let C_1 and C_2 be discharged at the time $t = 0$ ($Q_0 = 0$). Between the terminals there is a voltage source which supplies a constant current I. The same current I flows through C_1 and C_2, therefore both capacitors have the same charge $Q(t')$ at the time t'. The voltage is determined by Eq. 3.12 and the mesh rule (Eq. 3.9).

$$U_1(t') = \frac{Q(t')}{C_1} \quad U_2(t') = \frac{Q(t')}{C_2}$$
$$U(t') = U_1(t') + U_2(t')$$

The following therefore applies to the replacement capacity C.

$$\frac{1}{C} = \frac{1}{C_1} + \frac{1}{C_2}$$
$$C = \frac{C_1 \cdot C_2}{C_1 + C_2}$$

This formula can also be extended to any number of capacitors.

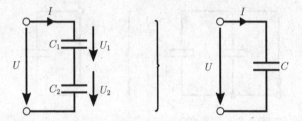

Fig. 3.14 Capacitors connected in series and the equivalent circuit diagram

$$\frac{1}{C_{tot}} = \sum_{i=1}^{N} \frac{1}{C_i} \tag{3.17}$$

Thus the total capacitance of capacitors connected in parallel behaves in the same way as the total resistance of resistors connected in series and vice versa.

3.4.4 Impedance

As can be seen from formula 3.14, the current and voltage curves for the capacitor are dependent on time. This makes it difficult to calculate networks containing capacitors easily, especially when no DC voltage is present. *Impedance* was introduced so that the methods of analysis for resistance networks can still be used. It describes the frequency-dependent, complex resistance of capacitors and coils, which can be used mathematically like an ohmic (normal) resistance. The impedance of a real capacitor is

$$\underline{Z}_C = R + \frac{1}{j\omega C} \tag{3.18}$$

Here, ω represents the angular frequency of the applied current/voltage and R the parasitic resistance in real components. For ideal capacitors, which are usually used for calculations, the impedance is thus

$$\underline{Z}_C = \frac{1}{j\omega C} \tag{3.19}$$

The "j" is a special feature of electrical engineering. Since i usually designates a current, its use as an imaginary unit could lead to confusion. In electrical engineering equations, $j = \sqrt{-1}$ is therefore used. To illustrate this, let us take the voltage divider in Fig. 3.15 as an example.

Using impedance, the voltage divider rule from Eq. 3.5 can also be used here if U(t) is an AC voltage.

$$U_C(t) = U(t) \cdot \frac{\underline{Z}_C}{\underline{Z}_C + R}$$

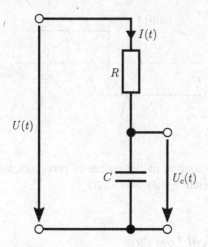

Fig. 3.15 Complex voltage divider

3.5 Switching Behavior of RC Elements

Circuits consisting of capacitors and resistors (and coils) are of interest for practical purposes. The simplest of such quadripoles are the first-order low and high pass, each consisting of a capacitor (C) and a resistor (R) (see Fig. 3.16). These circuits are widely used for filtering or selecting signals (see also Sects. 6.2.3.2, 6.2.3, 7.2 and 7.3.2). They can also be used as a model for the transfer function of electrical lines (see also Sect. 11.4.1).

Since these circuits are therefore of great relevance, it is worth taking a closer look. The behavior of these RC elements during switching operations which occur particularly in digital technology, is considered below. To analyze this behavior of four-terminal networks, jump functions (see Fig. 3.17) are often used.

$$U_{in}(t) = \begin{cases} U_1 & \text{for } t < t_0 \\ U_2 & \text{for } t \geq t_0 \end{cases} \tag{3.20}$$

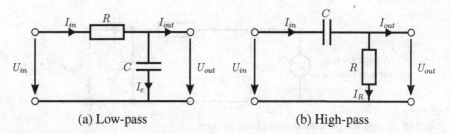

(a) Low-pass   (b) High-pass

Fig. 3.16 RC elements

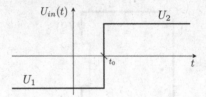

Fig. 3.17 Jump function

These functions are used to model switch-on processes, but can also serve as a simplified model of digital signals (see 5.2.5).

3.5.1 The Electrical Low Pass

A *low pass* is a circuit that filters out frequencies greater than a cut-off frequency f_{co}. With an ideal low pass, this would mean that all signal components with frequencies below the cut-off frequency would be passed on unchanged, while all signal components with higher frequencies would be completely suppressed. In reality, the lower frequencies are still affected, while the higher frequencies are not completely hidden. In particular, frequencies that are only slightly above the cut-off frequency are usually only slightly affected and their suppression increases as the frequency rises. Since no ideal low pass exists, the real switch-on and switch-off times must be known in order to assess the behavior of a real low pass correctly. First, the switch-on process is considered. If

$$U_{in}(t) = \begin{cases} 0 & \text{for } t < t_0 \\ U_0 & \text{for } t \geq t_0 \end{cases}$$

and $U_{out} = 0$ for $t \leq t_0$, i.e. the capacitor is uncharged.

By applying the nodal rule (Eq. 3.8) to the circuit in Fig. 3.18, the following results are obtained

$$I_{in} = I_C + I_{out}$$

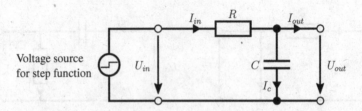

Fig. 3.18 Simple low pass circuit with a jump function

No current flows from the open terminals in the figure without a connected load. Therefore, $I_{out} = 0$ applies. With the mesh rule (Eq. 3.9), it follows

$$I_{in} = I_C$$
$$U_{in} = I_C \cdot R + U_{out} \quad \text{(mesh rule)}$$

From Eq. 3.12 it follows that $U_{out} = \frac{Q}{C}$ and thus

$$U_{in} = I_C \cdot R + \frac{Q}{C}$$
$$U_{in} = I_C \cdot R + \frac{1}{C} \int_{t_0}^{t} I_C dt$$

Derivation of the equation results in

$$\frac{dU_{in}}{dt} = R \cdot \frac{dI_C}{dt} + \frac{I_C}{C} \quad \text{(low pass equation)}$$

The jump function results in $U_{in} = U_0$ being constant for $t \geq t_0$, thus yielding

$$0 = R \cdot \frac{dI_C}{dt} + \frac{I_C}{C}$$

A function that satisfies this condition (determined experimentally) is

$$I_C(t) = I_0 \cdot e^{-\frac{t-t_0}{RC}}$$

Because of the derivation
$$\frac{dI_C}{dt} = -\frac{I_0}{RC} \cdot e^{-\frac{t-t_0}{RC}}$$

Insertion in the low-pass equation for $U_{in} = U_0 = const$ finally gives

$$0 = -\frac{R \cdot I_0}{RC} \cdot e^{-\frac{t-t_0}{RC}} + \frac{I_0}{C} \cdot e^{-\frac{t-t_0}{RC}}$$
$$= I_0 \cdot e^{-\frac{t-t_0}{RC}} \left(-\frac{R}{RC} + \frac{1}{C} \right)$$

To determine I_0, the initial condition is used: $U_{out} = 0$ for $t = t_0$. For $t = t_0$ the mesh equation $U_{in} = I_C \cdot R + U_{out}$ therefore gives

$$U_{in} = U_0 = I_C \cdot R + 0 = I_0 \cdot R \cdot e^{-\frac{t-t_0}{RC}} \bigg|_{t=t_0} = I_0 \cdot R$$

$$\Rightarrow I_0 = \frac{U_0}{R}$$

At time t_0 the capacitor therefore has no resistance for a short time and behaves like a short-circuit. This means that, for an infinitesimal period of time, the current through the capacitor becomes mathematically infinitely large and the voltage dropping across the capacitor falls to zero.

For *switch-on processes* in the low pass in general, the current I_C becomes

$$I_C = \frac{U_0}{R} \cdot e^{-\frac{t-t_0}{RC}} . \tag{3.21}$$

And thus for the mesh equation

$$U_{out} = U_0 \cdot \left(1 - e^{-\frac{t-t_0}{RC}} \right) \tag{3.22}$$

You also often come across the notation

$$U_{out} = U_0 \cdot \left(1 - e^{-\frac{t-t_0}{\tau}} \right) \text{ and } I_C = \frac{U_0}{R} \cdot e^{-\frac{t-t_0}{\tau}} \tag{3.23}$$

Here, $\tau = R \cdot C$ is the so-called *time constant* of the RC element, since it contains all the information needed to estimate the time response of the circuit.

3.5.1.1   Low Pass Switch-off Process

In the same way as the switch-on process, the switch-off process and the resulting voltage curve can be calculated.

For U_{in} a jump function is used again, as in Fig. 3.19, with

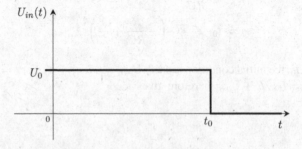

Fig. 3.19 Jump function for the low pass switch-off process

$$U_{in}(t) = \begin{cases} U_0 & \text{for } t < t_0 \\ 0 & \text{for } t \geq t_0 \end{cases}$$

If $U_{out} = U_{out0}$ for $t = t_0$, then for $t \geq t_0$ it follows

$$U_{out}(t) = U_{out0} + \frac{1}{C} \int_{t_0}^{t} I_C(t) dt$$

Insertion into the familiar mesh equation $U_{in}(t) = I_C(t) \cdot R + U_{out}(t)$ then results in

$$U_{in}(t) = I_C(t) \cdot R + \frac{1}{C} \int_{t_0}^{t} I_C(t) dt + U_{out0}$$

At the switch-off time $t = t_0$ the jump function yields $U_{in}(t) = 0$ and therefore

$$U_{in}(t_0) = I_C(t_0) \cdot R + 0 + U_{out0}$$
$$0 = I_0 \cdot R + U_{a0} \quad \text{with } I_0 = I_C(t_0)$$
$$\Rightarrow I_0 = -\frac{U_{out0}}{R}$$

Insertion into the known solution for $I_C(t)$ for the switch-off process gives

$$I_C(t) = -\frac{U_{out0}}{R} \cdot e^{-\frac{t-t_0}{RC}} \tag{3.24}$$

Insertion into the mesh equation with $U_{in}(t) = 0$ for $t \geq t_0$ yields

$$U_{out}(t) = U_{in}(t) - I_C(t) \cdot R$$
$$= U_{out0} \cdot e^{-\frac{t-t_0}{RC}} \tag{3.25}$$

3.5.1.2 Switching Process in the Low Pass

From the equations for the process of switching on and off, a general equation for switching processes in the low pass can be determined. For the general jump function

$$U_{in}(t) = \begin{cases} U_1 & \text{for } t < t_0 \\ U_2 & \text{for } t \geq t_0 \end{cases}$$

with $U_{out}(t_0) = U_{out0}$ and the mesh equation for $t = t_0$, it applies

$$U_2 = I_0 \cdot R + U_{out0}$$
$$\Rightarrow I_0 = \frac{U_2 - U_{out0}}{R}$$

It follows

$$I_C(t) = \frac{U_2 - U_{out0}}{R} \cdot e^{-\frac{t-t_0}{RC}} \tag{3.26}$$

$$U_{out}(t) = U_2 - (U_2 - U_{out0}) \cdot e^{-\frac{t-t_0}{RC}} \quad (\text{for } t \geq t_0) \tag{3.27}$$

Thus it becomes apparent that, as can be seen in Fig. 3.20, when switching the input of a low-pass circuit, the output does not follow the input directly, but the signal is "looped", i.e. the high frequency components are filtered. If one now imagines an AC voltage at the input, which alternates between a positive and a negative voltage with increasing frequency (corresponds in Fig. 3.20 to a t_0 moving to the left), one arrives at a point at which the capacitor no longer charges quickly enough to reach the maximum voltage, and/or no longer discharges quickly enough to accept the negative peak voltage.

The higher the frequency, the clearer the effect that the amplitude of the output voltage becomes lower. At an infinitely high frequency, the output of the capacitor therefore remains on the 0 V-line, i.e. it behaves like a short-circuit, while at low frequencies, the delay of the switching points is no longer noticeable. Here the capacitor behaves almost like an open terminal. This behavior gives the low pass its name.

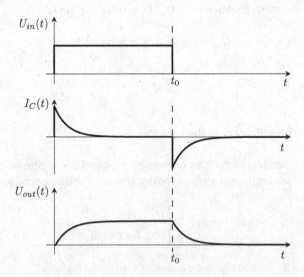

Fig. 3.20 Current and voltage curve at the low pass for switching-on and -off processes

3.5.2 Switching Behavior of a High-Pass Filter

The position of the capacitor and resistor are now switched, as shown in Fig. 3.21, resulting in the latter being parallel to the output voltage. The analysis of the general switching process is carried out in the same way as with the low-pass filter. The figure above is used for this purpose. Since no current can flow off when the terminal is open, the nodal rule gives

$$I_{in} = I_R$$

Applying the mesh equation then gives $U_{in}(t) = U_C(t) + U_{out}(t)$. With $U_{out}(t) = I_R(t) \cdot R$ and $U_C = \frac{Q}{C}$ it follows

$$U_{in}(t) = \frac{Q(t)}{C} + I_R(t) \cdot R$$

$$= \frac{1}{C} \int_{t_0}^{t} I_R(t)dt + \frac{Q_0}{C} + I_R(t) \cdot R$$

with the charge Q_0 of the capacitor at time t_0.

By derivation (with $U_{in} = const$ for $t \geq t_0$), it follows

$$0 = \frac{I_R(t)}{C} + \frac{dI_R(t)}{dt} \cdot R$$

As with the low pass, an exponential equation results

$$I_R(t) = I_0 \cdot e^{-\frac{t-t_0}{\tau}} \quad \text{with } \tau = RC \qquad (3.28)$$

The initial condition for $t = t_0$ is

$$U_{in} = U_2 = \frac{Q_0}{C} + I_0 \cdot R \cdot e^0$$

$$\Rightarrow I_0 = \frac{U_2 - U_{C0}}{R} \quad \text{with } U_{C0} = \frac{Q_0}{C}$$

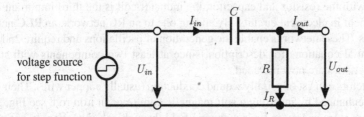

Fig. 3.21 Simple high-pass filter circuit with jump function

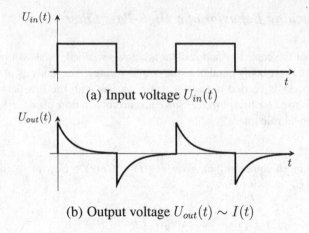

(a) Input voltage $U_{in}(t)$

(b) Output voltage $U_{out}(t) \sim I(t)$

Fig. 3.22 Voltage curve at the high pass during the switching-on and -off processes

This gives the general solution

$$I_R(t) = \frac{1}{R}(U_2 - U_{C0})e^{-\frac{t-t_0}{\tau}} \tag{3.29}$$

$$U_{out}(t) = I_R(t) \cdot R \tag{3.30}$$

$$U_{out}(t) = (U_2 - U_{C0})e^{-\frac{t-t_0}{\tau}} \tag{3.31}$$

In contrast to the low-pass filter, the output voltage for the high-pass filter is proportional to the current (see Fig. 3.22). Thus the high frequencies are now passed on to the output, while low frequencies are suppressed.

3.6   Resistor-Capacitor-Inductor Networks

3.6.1   Inductor

Along with the resistor and capacitor, the inductor/coil is the third important linear component in electrical circuits. By adding one to an RC network, an RLC network emerges. These networks enable the generation of oscillations and require 2nd order differential equations for description, since at least two components with storage characteristics are now included.

Inductors consist of spirally wound conductors (usually copper wire). Their effect can be enhanced by inserting a soft magnetic core (e.g. an iron rod, see Fig. 3.23). Inductors store energy in their magnetic field through *self-induction*, in contrast to capacitors, which store energy in an electric field between their plates. Due to the mechanical complexity of their production and the materials used, inductors are

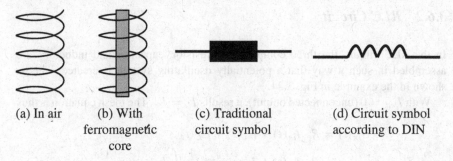

(a) In air   (b) With   (c) Traditional   (d) Circuit symbol
                     ferromagnetic   circuit symbol   according to DIN
                     core

Fig. 3.23 Types and graphical symbols of inductors

relatively expensive compared to other components, which is why they are used less often than, for example, capacitors and resistors. They are used, for example, in oscillators, filter circuits, switching power supplies and in modeling of side effects in circuits and transmission lines. The most important parameter for inductors is their inductance L, also referred to as inductivity.

$$L = \frac{N^2 \cdot \mu \cdot A}{l} \qquad (3.32)$$

with the number of turns of the inductor N, the permeability μ (magnetic conductivity of the material) in Vs/Am, the cross-sectional area of the coil A and the field line length (corresponds to the length of the coil) l. The unit of inductance is the *Henry* [H], where 1 H = 1Vs/A, named after Joseph Henry, who invented the electromagnetic relay in 1835.

An *inductive voltage drop* U_L occurs through the inductor due to self-induction.

$$U_L = L \cdot \frac{dI}{dt} \qquad (3.33)$$

3.6.1.1 Impedance of the Inductor

As for the capacitor, there is also a formula for the impedance of an inductor. The general case is

$$\underline{Z}_L = R + j\omega L \qquad (3.34)$$

And for an ideal inductor

$$\underline{Z}_L = j\omega L \qquad (3.35)$$

3.6.2   RLC Circuit

In the RLC circuit, the three components—resistor, capacitor and inductor—are assembled in such a way that a potentially oscillating system is created. This is shown in the example in Fig. 3.24.

With $I_{out} = 0$ (unconnected output), it results $I_L = I_{in}$. The mesh equation is thus

$$U_{in}(t) = R \cdot I_{in}(t) + U_L(t) + U_C(t)$$

$$= R \cdot I_{in}(t) + L \cdot \frac{dI_{in}}{dt} + \frac{1}{C} \int I_{in}(t)dt + \frac{Q_0}{C}$$

By derivation

$$\frac{dU_{in}}{dt} = R \cdot \frac{dI_{in}}{dt} + L \cdot \frac{d^2 I_{in}}{dt^2} + \frac{1}{C} \cdot I_{in}$$

This 2nd order linear differential equation is analytically solvable for

$$\frac{dU_{in}}{dt} = 0 \quad \text{(Input voltage constant)}$$

This applies to the considered jump function for $t \geq t_0$. Thus the following result can be derived.

$$I_{in} = I_0 \cdot e^{\left(-\frac{R}{2L} \pm \sqrt{\frac{R^2}{4L^2} - \frac{1}{LC}}\right) \cdot t} \tag{3.36}$$

To understand this equation, several cases must be considered depending on the exponent.

- **Case 1: Radicand < 0 (oscillation)**

$$I_{in} = I_0 \cdot e^{\left(-\frac{R}{2L} \pm j\sqrt{-\frac{R^2}{4L^2} + \frac{1}{LC}}\right) \cdot t} \tag{3.37}$$

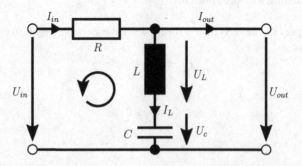

Fig. 3.24 Example of an RLC circuit

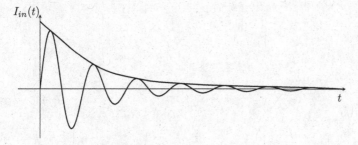

Fig. 3.25 Example of the oscillation of a RLC circuit with the decay behavior as envelope curve

If $\frac{R^2}{4L^2} - \frac{1}{LC}$ is negative, the root can only be solved in a complex way. Thus, as can be seen in Eq. 3.37, it is the imaginary part of a complex number which is in the exponent of the equation. The solution for I_{in} therefore has the general form

$$I_{in} = I_0 \cdot e^{(\alpha \pm j\beta)t} = I_0 \cdot e^{\alpha t} \cdot e^{\pm j\beta t}$$

According to Euler's equation

$$e^{\pm j\beta t} = \cos(\pm\beta t) + j\sin(\pm\beta t)$$
$$\Rightarrow I_{in} = I_0 \cdot e^{\alpha t} \cdot (\cos(\pm\beta t) + j\sin(\pm\beta t))$$

A real current measurement shows only the real part of this complex solution. So every solution of the equation contains an oscillation, as can be seen in Fig. 3.25.

- **Case 2: Radicand ≥ 0**

However, if the radicand is greater than 0, the exponent has a purely real value. Thus the complex solution is omitted and the electrical circuit is *free of oscillation*. This is also referred to as being *overdamped*. The special case radicand $= 0$ is called *critically damped*, because only just no oscillation occurs.

By changing the dimensions of the components of this circuit, oscillations of different frequency, phase and amplitude can be generated. This is used, for example, for the signal generators mentioned in Sect. 2.8. Many microcontrollers also use RLC circuits as internal clock generators when the precision requirements of the clock are not too high.

Chapter 4
Basic Electronic Circuits

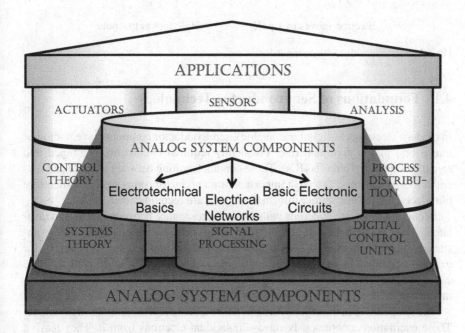

This chapter introduces the basics of electronics. In addition to a basic understanding of the functional principle of semiconductors, the introduction of the most important semiconductor components, such as diodes and transistors, provides the opportunity to investigate complex circuits. These play an important role, for example, in the conversion of analog signals into digital signals (see Sect. 7.5) and in the power electronics of electric motors (see Sect. 9.2). Furthermore, amplifier circuits and simple logic circuits are introduced on this basis. Finally, the operational amplifier provides the basis for mapping mathematical functions such as addition, subtraction and integration in analog circuits.

© The Author(s), under exclusive license to Springer Nature Switzerland AG 2021   45
K. Berns et al., *Technical Foundations of Embedded Systems*, Lecture Notes in Electrical
Engineering 732, https://doi.org/10.1007/978-3-030-65157-2_4

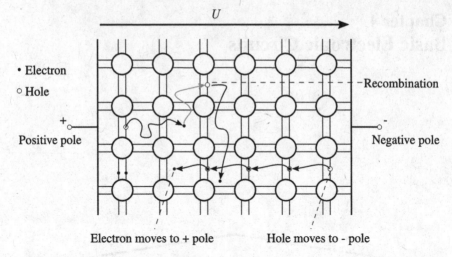

Fig. 4.1 Si crystal with a voltage applied

4.1 Foundations of Semiconductor Technology

In modern electronics (digital and analog), semiconductor components (e.g. diodes, transistors) play a dominant role. Frequently required circuits are combined into so-called integrated circuits (ICs). These contain anything between a single and several billion transistors. The most important semiconductor components are therefore introduced below. For further information [1–3] are recommended.

Semiconductors are materials with a specific resistance between metals and insulators. They belong to the 4th main group of the periodic table. This includes substances such as silicon (Si), germanium (Ge) and gallium arsenide (GaAs). If, for example, completely regular crystal structures are formed from silicon, they are called *monocrystals*. Since all electrons are bound within the monocrystal, a silicon monocrystal is an insulator, apart from its *intrinsic conduction*.

The temperature of the crystal corresponds to mechanical oscillations in the lattice. These oscillations continuously release individual electrons from it. They leave a positively charged *hole* in the crystal lattice. So, there are always electron-hole pairs (*pair formation*).

When a free electron approaches a hole, *recombination* may occur due to electrical attraction. A hole catches a free electron again. When a voltage is applied to the Si crystal lattice, the (*thermally*) free electrons act as negative charge carriers, the holes as positive charge carriers (Fig. 4.1), so theoretically an electric current can flow. At room temperature, however, the intrinsic conduction is almost negligible.

4.2 Doping of Semiconductors

In addition to the intrinsic conduction of a semiconductor mentioned above, the number of free charge carriers and thus their conductivity can be increased by introducing foreign atoms (doping) into the pure semiconductor crystal. A distinction is made between two types of doping: *n-doping* and *p-doping*. N-doping is carried out with pentavalent elements, e.g. phosphorus, and is characterized by the excess *negative* charge carriers introduced by the foreign atom—p-doping uses trivalent substances, e.g. aluminum, and is characterized by the *positive* hole which results from doping.

Doping thus increases the number of free charge carriers. The number of additional *majority carriers* (holes for p-doping, electrons for n-doping) is equal to the number of foreign atoms and is temperature-independent. The number of additional *minority carriers* (electrons for p-doping, holes for n-doping) is temperature-dependent, but only plays a role at transitions between p-doped and n-doped semiconductors.

Note The specific resistances of the doped semiconductors ($4 \cdot 10^{-1}$ Ωm - $6 \cdot 10^5$ Ωm) achieved are relatively high compared to copper ($18 \cdot 10^{-9}$ Ωm). I.e. cable lengths must be kept short.

4.3 The pn-Transition

A p-conductor and an n-conductor are now to be joined by melting. Before joining, both sides are electrically neutral. The n-side has a high concentration of electrons and the p side a high concentration of holes, i.e. a low concentration of electrons.

Due to heat movement, different concentrations try to balance each other out ($\rightarrow$ diffusion). This happens when the n- and p-conductors are now joined. In the process, a diffusion current of electrons first flows from the n-conductor into the p-conductor. The diffusion current quickly comes to a standstill because the boundary layer is electrically charged. Positive ions remain on the n side. The diffused electrons are relatively firmly embedded in the holes on the p side, resulting in a negative excess charge. Their electric field pushes back further electrons diffusing from the n-side. The result is a depletion layer, "insurmountable" for electrons, without freely movable charge carriers. All free electrons and holes are recombined (see Fig. 4.2).

On both sides of the depletion layer there are now separate charges opposite each other, as in a capacitor. The charges generate an electric field and thus a voltage U_D. This *diffusion voltage* is relatively small and depends on the doping and the temperature. It is about 0.7 V for silicon at room temperature and 0.3 V for germanium. Figure 4.3 shows the relationship between the diffusion voltage and the pn junction.

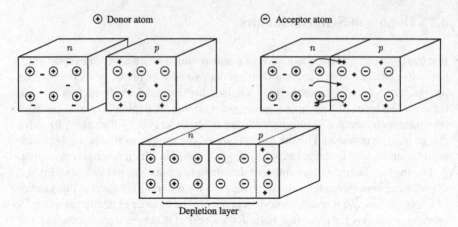

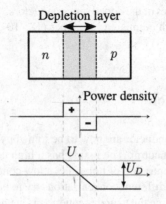

Fig. 4.2 Combination of p-doped and n-doped material

Fig. 4.3 Equilibrium between recombination and diffusion voltage U_D

4.4 Diodes

The electronic component consisting of the assembled p- and n-conductors is called a *diode*. It behaves like a check valve for electric current.

In Fig. 4.4, a diode is electrically connected to a voltage source and a lamp. On the left of the figure, the pn-junction is wired in the *forward direction*. The external voltage U_F "pushes" the electrons of the n-conductor first into the depletion layer and then, if U_F is greater than the diffusion voltage, over the depletion layer so that they can flow off through the p-conductor. The result is a current flow that causes the lamp to light up. The circuit symbol for the diode is shown above the circuit. The diode voltage U_F is approximately equal to the diffusion voltage when connected in the forward direction, since the outer zones of the diode represent a negligible resistance. It is (almost) independent of the battery voltage, i.e. most of the battery voltage drops through the lamp: $U_L = U_B - U_F$. The electric current

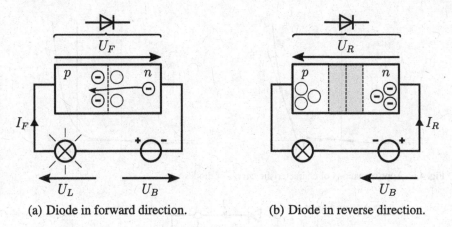

(a) Diode in forward direction.   (b) Diode in reverse direction.

Fig. 4.4 Simple electronic circuit with a diode acting as check valve for the current

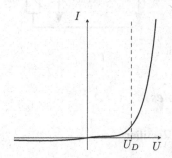

Fig. 4.5 Characteristic curve of a diode

through the diode grows exponentially with the applied voltage U_F: $I_F \sim e^{U_F}$, as shown in Fig. 4.5.

If the diode is connected in the *reverse direction*, i.e. the current direction is reversed as on the right of Fig. 4.4, the free charge carriers are pulled away from the boundary layer. This makes the carrier-free zone, i.e. the depletion layer, wider. The diode blocks the current ($U_R = U_B$). Only a negligibly small reverse current I_R flows due to thermal pair formation. For silicon this is about 1 nA and for germanium about $10\,\mu A$.

In practice, the diode characteristic curve can usually be approximated using the threshold voltage $U_{th} \approx U_D$ (application-dependent) (see Fig. 4.6). The diode is therefore suitable for reverse polarity protection, but also for overvoltage protection, for rectifying signals and for some other areas of application, which are considered below. An overview of common diode types can be found in Appendix B.4.

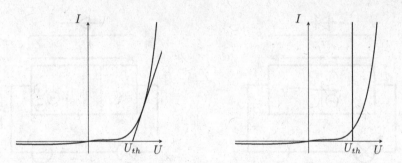

Fig. 4.6 Approximations of characteristic curves of diodes

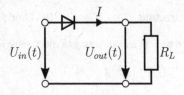

Fig. 4.7 Simple rectifier circuit

4.4.1 Wiring Examples

4.4.1.1 The Diode as a Rectifier

Diodes can be used to build simple rectifier circuits. The purpose of rectifiers is to convert an AC voltage (or voltage with alternating current components) into a DC voltage. This is a commonly used circuit to power an embedded system that is operated by a generator or the mains. Since the embedded system usually requires a DC voltage but the energy source in this case only supplies an AC voltage, a transformation must take place. The circuit in Fig. 4.7 takes advantage of the fact that diodes conduct current only in their forward direction, so that the use of a diode directly blocks the current in the opposite direction, as occurs with AC voltages in every second half-wave, see Fig. 4.8.

The simplified diode characteristic curve is assumed, i.e. for $U_{in}(t) > U_{th}$ the diode is conductive with a voltage drop of U_{th}. It follows

$$U_{out}(t) = \begin{cases} U_{in}(t) - U_{th} & \text{for } U_{in} > U_{th} \\ 0 & \text{for } U_{in} \leq U_{th} \end{cases} \tag{4.1}$$

4.4.1.2 Rectifier Circuit with Smoothing Capacitor

A constant output voltage is required for the generation of DC voltages to supply power to electronic devices. A rectifier circuit with a *smoothing capacitor* can offer

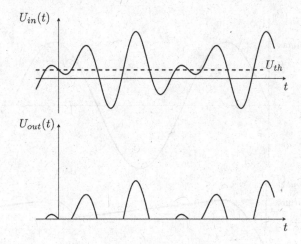

Fig. 4.8 Signal course of the rectification using a diode

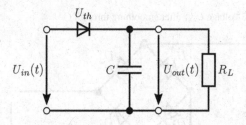

Fig. 4.9 Rectifier circuit with smoothing capacitor

this, as shown in Fig. 4.9. It makes use of the fact that the capacitor cannot charge up infinitely quickly in connection with an (assumed) ohmic resistance in the load, which filters out higher-frequency voltage changes.

The capacitor is charged up to the peak value $U_{out\text{max}} = U_{in\text{max}} - U_{th}$ and discharged via R. As soon as $U_{in} > U_{out} + U_{th}$ applies, the diode becomes conductive again and C is charged up to $U_{out\text{max}}$. A *rippled* DC voltage U_{out} is generated (see Fig. 4.10).

$$U_{out}(t) = U_{out\text{max}} \cdot e^{-\frac{t}{\tau}} \tag{4.2}$$

$$\text{with } \tau = C \cdot R_L \text{ (time constant)} \tag{4.3}$$

For AC input voltage $U_{in}(t)$ with a period T, $\tau >> T$ must apply so that the ripple is small. The ripple is characterized by the peak-to-peak voltage $U_{PP} = U_{out\text{max}} - U_{out\text{min}}$.

The simple rectifier only uses every second half-wave (*half-wave rectifier*), a more favourable approach is given with the *bridge rectifier* (see Fig. 4.11). The bridge

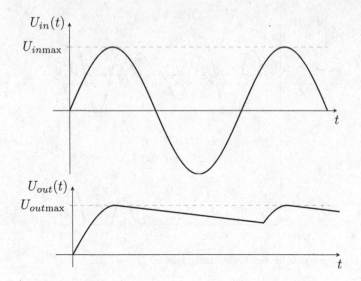

Fig. 4.10 Rippled DC voltage U_{out} after smoothing through C

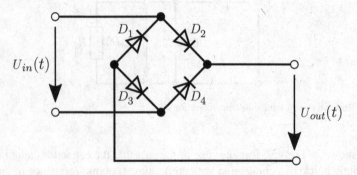

Fig. 4.11 Circuit diagram of a bridge rectifier

rectifier utilizes each half-wave, which is made possible by charging the smoothing capacitor twice per period. However, the switching voltage losses add up to $2 \cdot U_{th}$.

Germanium diodes are therefore used in practice, as they have a much lower switching voltage of $U_{th} = 0.25$ V compared to silicon diodes with $U_{th} = 0.7$ V. The signal curve is shown in Fig. 4.12.

$$U_{out}(t) = \begin{cases} |U_{in}(t)| - 2 \cdot U_{th} & \text{for } |U_{in}| > 2 \cdot U_{th} \\ 0 & \text{for } |U_{in}| \leq 2 \cdot U_{th} \end{cases} \qquad (4.4)$$

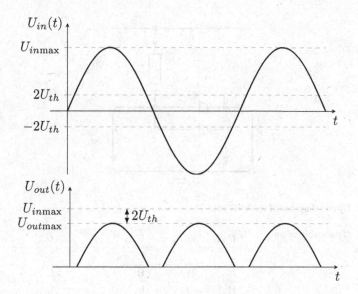

Fig. 4.12 Signal curve at a bridge rectifier without smoothing capacitor

4.4.2 Diodes as Switches

Digital systems are mainly based on logical switching elements for the representation of binary states, so-called *logical gates*. For example, modern data storage devices, many special ICs and even processors consist largely of an interconnection of such elements. In principle, only two electrical voltages are distinguished at the output, one *HIGH* and one *LOW* (usually GND). These voltages are assigned *logically 1* or *logically 0* (e.g. HIGH corresponds to 1, LOW corresponds to 0). In DTL circuit technology (diode transistor logic), diodes are used to create such logic gates.

4.4.2.1 AND Gate

The AND gate is a circuit with (at least) two inputs, the output of which only assumes logic 1 if all inputs were also logic 1. According to the above assignment, this means for the circuit shown in Fig. 4.13, that if at least one input line is LOW, then the corresponding diode is conductive. $U_{out} = U_{th}$ (LOW) applies. This is explained in detail in the truth table in Table 4.1. If both inputs are $U_{in1} = U_{in2} = U_b$ (HIGH), there is no connection from U_b to the ground via the diodes. Then $U_{out} = U_b$ is (HIGH).

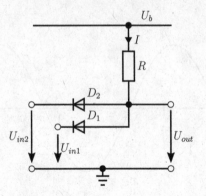

Fig. 4.13 Circuit diagram of an AND gate

Table 4.1 Truth table of an AND gate

U_{in1}	U_{in2}	U_{out}
L	L	L
L	H	L
H	L	L
H	H	H

Table 4.2 Truth table of an OR gate

U_{in1}	U_{in2}	U_{out}
L	L	L
L	H	H
H	L	H
H	H	H

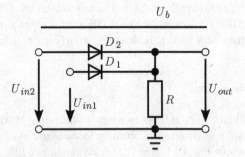

Fig. 4.14 Circuit diagram of an OR gate

4.4.2.2   OR Gate

With the OR gate, the output assumes logic 1 if at least one of the inputs is also logic 1 (see Table 4.2). As long as for the circuit in Fig. 4.14 $U_{in1} = U_{in2} = 0$ V

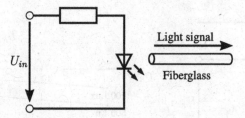

Fig. 4.15 LED as data transmitter

applies, $U_{out} = 0$ V is (LOW). As soon as at least one of the inputs is equal to U_b, a current flows through the corresponding diode(s) and the following applies: $U_{out} = U_b - U_{th}$ (HIGH).

Diodes can be used to build AND and OR gates, the diode can thus be used as a switch. However, NOT gates are not possible, which is why not all circuits can be built with diodes.

4.4.3 Special Diodes

In addition to the basic diodes presented so far, various special diodes have been developed which are used, for example, for applications in power electronics or as surge protection. Of particular importance for the development of embedded systems are the light-emitting and photodiodes presented below.

During recombination with holes, electrons emit energy in the form of heat or light. The light yield can be increased and the adjustment of certain wavelengths made possible by suitable selection of semiconductor materials. *Light emitting diodes* (*LED*) emit light when current flows through them in the forward direction.

LEDs are used for instrument displays, messaging, optocouplers, laser diodes and, of course, as lighting fixtures. As can be seen in Fig. 4.15, an electrical signal (shown here as a voltage source) can be converted into a light signal by means of an LED with a simple circuit, which is transmitted through a glass fiber line. The fact that an LED, in contrast to e.g. a light bulb, does not need any warm-up time is exploited here and explains why rapid changes between light and dark can be achieved.

In order to convert the light signal back into an electrical signal on the receiving side, another component is required, the *photodiode*.

Photodiodes are used to receive light signals. In addition to the thermal pairs, incidence of light on the barrier layer generates free charge carriers by pair formation. This increases the conductivity of the diode. If they are operated in reverse direction, the flowing reverse current is therefore proportional to the illuminance (cf. Fig. 4.16).

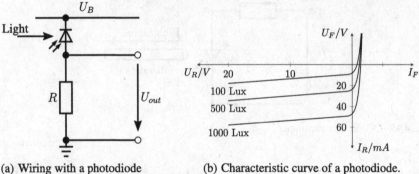

(a) Wiring with a photodiode   (b) Characteristic curve of a photodiode.
$U_{out} \sim$ illuminance.

Fig. 4.16 Characteristic curve and wiring of a photodiode

4.5 Bipolar Transistor

The second important semiconductor component is the transistor. In practice, it is often the case that a received electrical signal is only very weak, i.e. has only a small amplitude difference or a weak current strength. In order to use this signal effectively, it must be amplified. Transistors form the basis for this in analog and sensor technology (hifi amplifiers, residual light amplifiers, LIDAR preamplifiers, etc.). However, their greatest use is as switches in digital technology. While today powerless controllable *field effect transistors* (*MOSFETs*) are used, *bipolar transistors* that can bear larger currents are usually found in the analog area of embedded systems, for example, in many operational amplifiers (see below), or in analog signal processing. These bipolar transistors are first discussed before the structure and functionality of MOSFETs is explained in the next section.

In contrast to the diode, which consists of two different doping layers, the bipolar transistor consists of three alternating doping layers. These can be N-P-N layers or P-N-P layers. Accordingly, these transistors are called npn or pnp transistors (cf. Fig. 4.17). Only the npn type is discussed below. The pnp transistor functions in the same way but with reverse polarity.

The npn transistor consists of a highly N-doped emitter layer (E), a thin and weakly P-doped base layer (B) and a weakly N-doped collector layer (C). If the collector is positively polarized with respect to the base and the emitter, the *BC diode* (so the diode formed by the base layer and the collector layer) blocks the current.

The *BE diode* will also block as long as the *BE voltage* is below the threshold value. If the *BE voltage* increases above the threshold value, electrons flow from the emitter into the thin base. Due to the strong N-doping of the emitter, many electrons are drawn into the base by the *BE voltage*, of which only a few can flow off via the thin base. Most electrons diffuse into the *BC depletion area* and flow to the collector, driven by the positive *CE voltage*.

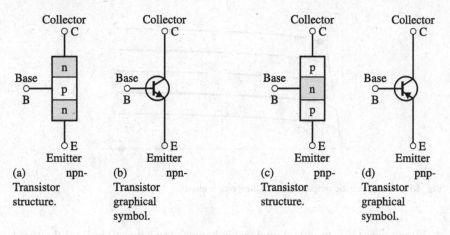

Fig. 4.17 Structure and graphical symbol of bipolar transistors

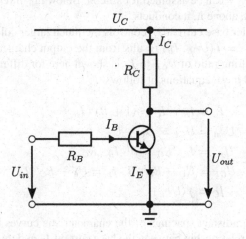

Fig. 4.18 Transistor as current amplifier

Two circuits result, as shown in Fig. 4.18. A small base current (input circuit) influences or controls a much larger collector current (output circuit). It is therefore said that the transistor amplifies the base current. The collector current I_C changes approximately in proportion to the base current I_B.

$$I_C = B \cdot I_B, \tag{4.5}$$

whereby the amplification factor B is typically about 250 (between 50 and 1,000). The *BE diode* in the input circuit behaves like the diode described above. The base current I_B—and thus also the collector current I_C—increases nearly exponentially with the input voltage U_{BE}.

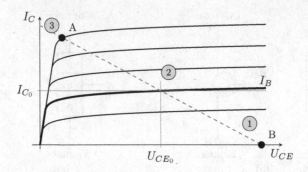

Fig. 4.19 Characteristic output curves of a bipolar transistor

If the transistor is used in digital technology as a switch, an idealized threshold voltage of about 0.7 V can be assumed for silicon. Below this threshold voltage, the transistor is locked, above it, it conducts.

The much smaller base current I_B controls the much larger collector current I_C. The relationship $I_C = f(U_{CE}, I_B)$ results from the output characteristics shown in Fig. 4.19. The resulting ratio of U_{CE} to I_C is shown here for different base currents I_B. From node and mesh equations, it follows

$$I_E = I_B + I_C = (1 + B) \cdot I_B \tag{4.6}$$

$$U_{out} = U_{CE} = U_C - U_{R_C} \tag{4.7}$$

$$U_{R_C} = I_C \cdot R_C = B \cdot I_B \cdot R_C \tag{4.8}$$

$$U_{CE} = U_C - R_C \cdot B \cdot I_B = U_C - R_C \cdot I_C \tag{4.9}$$

$$I_C = f(U_{CE}) \tag{4.10}$$

The (almost) equidistant spacings of the characteristic curves in Fig. 4.19 show the (almost) linear relationship between the base current I_B and the collector current I_C. The right area of the characteristics is the so-called working area (2). Here, I_C depends mainly on I_B and hardly at all on U_{CE} (horizontal characteristic curve). The left area is the saturation area (3). Here, U_{CE} is too small in proportion to I_B, so that $U_{CB} < 0$, which means, that the collector no longer functions as an "electron collector". The base in this area is saturated with charge carriers, so the magnification of I_B does not increase I_C. Only the diffusion pressure causes conductivity of the BC diode. In the cutoff area (1), I_B is so small that no current flows from the base and collector to the emitter.

As long as the input voltage change at the BE diode is small, the characteristic around the operating point (I_{C0}, U_{CE0}) can be regarded as approximately linear. This is called *small signal behavior* and allows the use of the equivalent circuit diagram in Fig. 4.20.

The differential resistances are defined as the gradient of the tangents at the operating point.

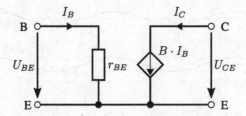

Fig. 4.20 Equivalent circuit diagram for small signal behavior

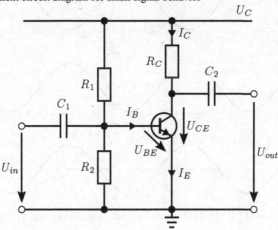

Fig. 4.21 Emitter circuit as AC voltage amplifier

$$r_{BE} = \frac{\partial U_{BE}}{\partial I_B}\bigg|_{U_{CE}=Const} \tag{4.11}$$

$$r_{BE} \approx \frac{U_T}{I_{B0}}, \; U_T \approx 26\,\text{mV} \tag{4.12}$$

4.5.1 Emitter Circuit

An important basic circuit for transistors is the *emitter circuit* shown in Fig. 4.21. It is used when small AC voltages have to be amplified or regulated, for example in audio amplifiers or transistor radios, and also in order to use the transistor as a switch. It forms the basis of the differential amplifiers and operational amplifiers introduced in Sect. 4.7.

A small AC voltage $U_{in}(t)$ in the millivolt range is applied to the input of the amplifier, as could be the case, for example, with a demodulated radio signal. This is to be amplified to operate a loudspeaker. The input voltage U_{in} is too low to drive the transistor directly, as it shuts off at voltages below 0.7 V (for silicon transistors). The voltage must therefore be increased by a direct component to an operating point OP, which is achieved by the voltage divider $R_1 - R_2$. The input voltage of the transistor now also fluctuates around the voltage $U_B \cdot R_2/(R_1 + R_2)$ at U_{in}.

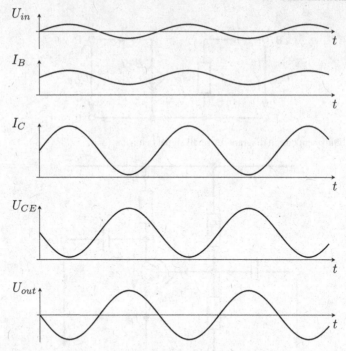

Fig. 4.22 Comparison of U_{in}, I_B, I_C, U_{CE} and U_{out} at the emitter circuit

Possible DC components in U_{in} are filtered out by C_1. The transistor now amplifies the mixed signal. At the operating point, the base current $I_B = I_{B0} + I_{B\sim}$ results from the current I_{B0} and the alternating current $I_{B\sim}$. If the operating point is selected correctly, the collector current I_C does not depend on U_{CE}, but only linearly on I_B. Since I_C flows through the collector resistor R_C and is limited by it, $U_{RC} \sim I_B$ and $U_{CE} = U_B - U_{RC} \sim I_B$ also apply. The capacitor C_2 filters out the direct component from this voltage, so that the output voltage U_{out} is proportional to U_{in}, but larger by the amplification factor of the transistor and rotated by 180°. This phase reversal is typical for this circuit type. The signals inside the circuit are compared in Fig. 4.22.

Like the base current, the collector-emitter voltage U_{CE} is composed of the voltage at the operating point U_{CE0} and the AC voltage $U_{CE\sim}$: $U_{CE} = U_{CE0} + U_{CE\sim}$. In order to determine both, the operating point setting and the AC voltage amplification are therefore considered individually below.

4.5.2 Operating Point

The calculation of the operating point is based on the equivalent circuit diagram in Fig. 4.23. The AC voltage source in Fig. 4.21 is set to $U_{in} = 0$. Since only the DC voltage source U_C is included in the system, capacitances are regarded as circuit

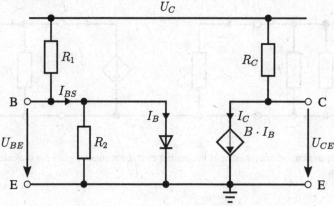

Fig. 4.23 Equivalent circuit diagram for the determination of the operating point

interruptions. The transistor itself can be considered as a diode and a *constant current source* coupled to its current. This is an abstract component that always keeps the impressed current (here: $B \cdot I_B$) constant. The DC current for base and emitter and the collector-emitter voltage U_{CE0} result to

$$I_{BS} = \frac{U_C - U_{BE}}{R_1}$$

$$I_{B0} = I_{BS} - \frac{U_{BE}}{R_2}$$

$$U_{CE0} = U_C - R_C \cdot I_{C0} = U_C - R_C \cdot B \cdot I_{B0} \tag{4.13}$$

R_1 and R_2 are now dimensioned in such a way that the output currents and voltages resulting from the known input currents and voltages lie within a range which is favorable for further processing. In particular, U_{BE} should be within the range of the transistor's threshold voltage, as otherwise overdrive or "underdrive" of individual half-waves of the output signal can result.

4.5.3 AC Voltage

To calculate the AC voltage, the supply voltage source is set to $U_C = 0$ and the equivalent circuit for small signals in Fig. 4.24 is used. As a replacement for the transistor, the input resistance r_{BE} and another constant current source are considered here. Now the input and output circuits are set up, the output voltage U_{out} and voltage amplification v are calculated. From the diagram in Fig. 4.25, $U_{in}(t) = 0.02 \sin(10\pi t)$ and the amplification $v \approx -20$ can be estimated. The resulting curves for $I_{B\sim}$, $I_{C\sim}$, U_{CE} and U_{BE} are summarized in Fig. 4.26.

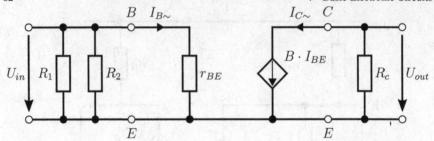

Fig. 4.24 Equivalent circuit diagram of the emitter circuit for small signals for the calculation of the AC voltage

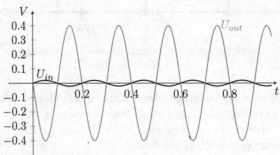

Fig. 4.25 Voltage curve of U_{in} and U_{out} at the emitter circuit

$$I_{B\sim} = \frac{U_{in}}{r_{BE}} \tag{4.14}$$

$$I_{C\sim} = B \cdot I_{B\sim} \tag{4.15}$$

$$U_{CE} = -I_{C\sim} \cdot R_C \tag{4.16}$$

$$U_{out} = U_{CE} = -\frac{U_{in} \cdot B \cdot R_C}{r_{BE}} \tag{4.17}$$

$$v = \frac{U_{out}}{U_{in}} = -\frac{B \cdot R_C}{r_{BE}} \tag{4.18}$$

4.6   Field Effect Transistors

Unipolar transistors (*field effect transistors, FET*) are particularly important in digital technology. The name results from the fact that, in contrast to the bipolar transistor, the amplification or conductivity of the transistor is not controlled by a current but by the strength of an applied voltage and the resulting electric field. This significantly reduces the current flow through the transistor and thus the power consumption,

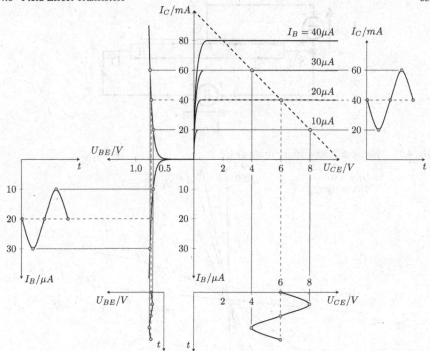

Fig. 4.26 Examples of characteristic curves of an emitter circuit. The diagram shows the dependencies between I_C and U_{CE} for different values of I_B, between I_C and U_{BE}, between U_{BE} and I_B and between U_{CE} and I_B. It also depicts the time courses for a sinusoidal change of the input

which is particularly advantageous for highly integrated circuits that can contain billions of transistors.

The basic principle is shown in Fig. 4.27. The n-doped semiconductor between the terminals D (for *drain*) and S (for *source*) is conductive in principle due to the excess of negative charge carriers. By applying a voltage U_{GS} to the insulated control electrode *gate* (G), an electric field is generated in the n-doped Si channel between drain and source. The free electrons from the channel are displaced, thus increasing the semiconductor resistance. This field effect transistor type is called *self-conducting* because it conducts at $U_{GS} = 0$.

Figure 4.28 shows the relationship of U_{DS} to I_D in a MOSFET for different values of U_{GS} for values in the millivolt/ampere range. It turns out that the transistor in this range functions as an approximately linear resistor that can be controlled by U_{GS}. In larger voltage ranges (see Fig. 4.29), it can be seen that this behavior only applies up to approximately $U_P = U_{GS} - U_T$. In the saturation range above this, however, it behaves similarly to the bipolar transistor (see Fig. 4.19). The main difference is that the amplification is not controlled by the input current, but by the voltage U_{GS}.

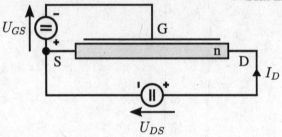

Fig. 4.27 Voltage driven switching using a FET

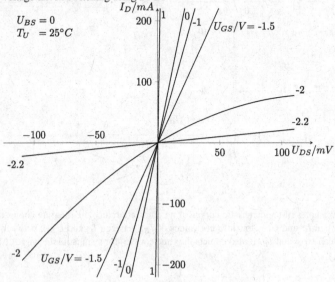

Fig. 4.28 Characteristic curves of a MOSFET for small U_{DS}

There are several types of field effect transistors, as shown in Fig. 4.30. All types are suitable as amplifiers. As digital switches, however, only self-locking n- and p-channel *MOSFETs (Metal Oxide Semiconductor Field Effect Transistor)* are significant, since their control voltage lies within the working range of the source-drain voltage. They will therefore be examined in more detail below.

4.6.1   Enhancement Mode MOSFETs

While the self-conducting NMOS transistor (n-doped MOSFET) requires both positive and negative voltages, self-locking, so called *enhancement mode* MOSFETs manage with only one voltage source. This makes it possible to build very compact circuits.

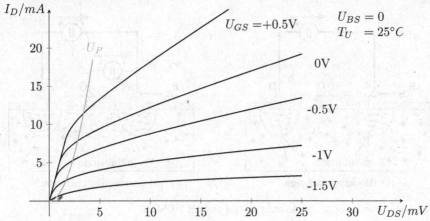

Fig. 4.29 The *Ohmic region* of the characteristic curves is only valid up to the *saturation voltage* $U_P = U_{GS} - U_T$; above: *saturation region*

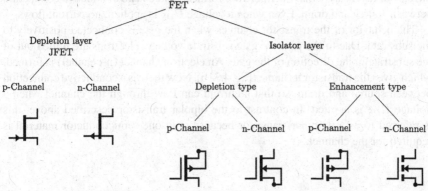

Fig. 4.30 Overview on types of FETs

The transistor again has three connections, which are called source, drain and gate (see Fig. 4.31). In addition, there is a bulk/body (B) substrate terminal which ensures that no charge carriers can flow off via the p-doped substrate in which the transistor is embedded. Two highly n-doped areas, source and drain, are inserted into this p-doped carrier. The third connection, the gate, is located above the connection area of the source and drain. This is a metal plate separated from the semiconductor substrate by an insulating layer (SiO_2). Substrate and gate connection thus form a kind of capacitor.

Due to the pn-junction between the substrate and the source/drain regions, the diodes shown in Fig. 4.31a are created. In addition, if the substrate terminal (B) has the lowest electrical potential, the diodes are polarized in the reverse direction and

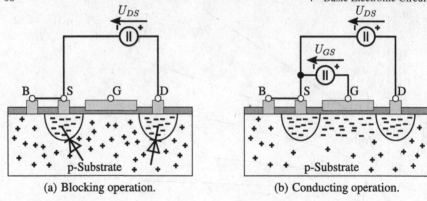

(a) Blocking operation.   (b) Conducting operation.

Fig. 4.31 Structure of an enhancement mode n-channel-MOSFET

the barrier layers shown in the figure are formed. There is no conductive connection between source and drain. Even when a voltage U_{DS} is applied, no current flows.

The behavior of the transistor changes when the gate is connected positively to the substrate. Due to the positive gate/substrate voltage, electrons are pulled out of the substrate in the direction of the gate. An electron channel (n-channel) is formed, which gives the transistor its name (Fig. 4.31b). Now there is a conductive connection between source and drain, so that a current can flow through the channel when a voltage U_{DS} is applied. In contrast to the bipolar transistor described above, this is referred to as a unipolar transistor because only one semiconductor material is required for the channel.

4.6.2 Depletion Mode MOSFETs

Figure 4.32 shows the structure of a depletion mode n-channel MOSFET. Here, too, the gate is separated from the substrate by a thin insulating layer. However, there is already a doped n-channel between drain and source. The stronger doping of the contact zones serves to prevent diode effects at the metal-semiconductor junctions. Charges can flow through the n-doped channel in an unconnected state. If a negative voltage is applied to the gate, the free charge carriers are displaced from this channel and the transistor becomes non-conductive.

This principle can also be reversed. Figure 4.33 shows the structure of a depletion mode p-channel MOSFET. Positive charge carriers can flow through the pre-doped p-channel. If a positive voltage is applied to the gate, these charge carriers are displaced (or electrons from the surrounding n-doped substrate are pulled into the holes) and the channel is thus blocked. The behavior is therefore almost identical with reverse polarity of all voltages. With the same degree of doping and the same dimensions,

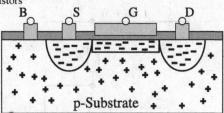

Fig. 4.32 Structure of a depletion mode n-channel MOSFETs

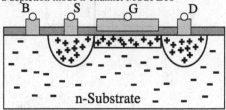

Fig. 4.33 Structure of a depletion mode p-channel-MOSFETs

the resistance of p-channel transistors is about 3 times higher than that of n-channel transistors, which is due to the lower mobility of the holes.

4.7 Basic Transistor Circuits

Transistors form the basis of all (micro-) electronic circuits. Some basic circuits are examined below. First the inverter is considered, which is the simplest digital circuit. The differential amplifier is then introduced as an extension of the emitter circuit, followed by the operational amplifier. The latter can be found in almost all analog-digital interfaces of embedded systems.

4.7.1 Transistor as an Inverter

The phase reversal described for the AC voltage amplifier in the emitter circuit (see Sect. 4.5) can also be used in the field of digital technology for inverting digital signals. The structure of a switch or inverter (Fig. 4.34) therefore looks very similar to this circuit, but the operating point setting can be dispensed with, since the input and output signals are of the same size and no negative voltages occur.

If the input voltage U_{in} is low, more precisely, less than 0.7 V (logical 0), the transistor blocks it. Therefore $I_C = 0$ and $U_{out} = U_B = 5$ V (logical 1). If the input voltage is high (logical 1), the transistor conducts and I_C thus rises to a high value. Due to this high collector current flowing through R_C, a high voltage U_{RC} drops across the collector, so that $U_{out} = U_B - U_{RC}$ is low (logical 0). The binary input signal is therefore inverted.

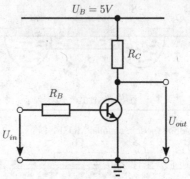

Fig. 4.34 Inverter circuit

When using the transistor as a switch, it is important that the transistor is not used in the amplification range. The transistor should either be definitively locked, i.e. $U_{in} << 0.7$ V or wide open (e.g. $U_{in} >> 2$ V), so that the output voltage is small enough and definitively logical 0 for a following logic gate. To ensure that these voltage intervals are maintained even with disturbed signals, a minimum signal-to-noise ratio must also be specified.

4.7.2   Differential Amplifier

The *differential amplifier* consists of two individual amplifiers T_1, T_2. The differential amplifier forms the heart of an operational amplifier. As its name suggests, the differential amplifier amplifies the difference between two input signals. If these two input voltages are changed evenly, the output of the amplifier does not change. Ideally, this so-called *common mode rejection* is zero.

Figure 4.35 shows the basic structure of a differential amplifier. These are two transistors which are coupled via their emitter and connected to a constant current source. In practice, this current source can be approximated more or less well by a simple transistor circuit or even by a common emitter resistor, but this reduces the quality of the amplifier. For the sake of simplicity, an ideal power source is considered for the analysis.

The symmetry of the components is important for the analysis and design of a differential amplifier. The two transistors T_1 and T_2 must have uniform characteristics and the same temperature. In the past, very high-quality, carefully selected and expensive components were necessary for this. The introduction of integrated circuits has greatly reduced this problem.

First, the operating point OP of the circuit is considered at $U_{in1} = U_{in2} = 0$ V. As a result of the second voltage source $U_{B-} < 0$ V, the BE diode of the two transistors is also conductive in this case, and the output voltages result to

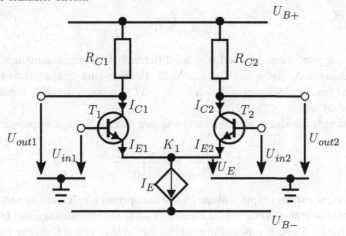

Fig. 4.35 Structure of a differential amplifier

$$U_{out1} = U_{B+} - I_{C1} \cdot R_{C1}$$
$$U_{out2} = U_{B+} - I_{C2} \cdot R_{C2} \rightarrow U_{out1_{OP}} = U_{out2_{OP}} = U_{B+} - I_{C_{OP}} \cdot R_C$$

This is based on the symmetry for $R_{C1} = R_{C2} = R_C$ and $I_{C1} = I_{C2} = I_C$.

In the node K_1, the emitter current I_E is composed of the two partial currents I_{E1} and I_{E2}. Since I_E is constant, I_{E1} and I_{E2} must always change as opposites.

$$\Delta I_{E1} = -\Delta I_{E2}$$

As $I_{C1/2} \approx I_{E1/2}$, the output voltages change accordingly.

$$\Delta U_{out1} \approx -\Delta I_{E1} \cdot R_C \text{ and} \tag{4.19}$$
$$\Delta U_{out2} \approx -\Delta I_{E2} \cdot R_C = \Delta I_{E1} \cdot R_C \tag{4.20}$$

and hence $\Delta U_{out1} = -\Delta U_{out2}$. Like the currents, the output voltages also change symmetrically.

The input voltages U_{in1} and U_{in2} are now increased evenly to a voltage U_{in} different from zero. Since the voltage drop across the BE diodes of the two transistors remains the same as the threshold voltage, the voltage across the current source also increases with U_{in}.

$$U_E = U_{in} - U_{BE}$$

Due to the constant current source and the symmetry of the two transistors both I_E and $I_{E1} = I_{E2} = I_E/2$ remain unchanged. This means that the collector currents I_{C1} and I_{C2} and, ultimately, the output voltages $U_{out1} = U_{out2} = U_{B+} - I_C \cdot R_C$ remain unchanged. Therefore, the common mode gain becomes

$$A_{cm} = 0$$

But what happens when U_{E1} and U_{E2} are different? Under the assumption that the voltage changes $\Delta U_{in1} = -\Delta U_{in2}$ are small, the transistor characteristics can be considered linear, which means that $\Delta I_{C1} = -\Delta I_{C2}$, since I_E remains constant due to the current source.

For a simple transistor circuit, it can be shown that the voltage amplification

$$A = \frac{dU_{out}}{dU_{in}} = \frac{U_B - U_{out}}{U_T} \tag{4.21}$$

depends on the current output voltage U_{out} at the operating point and on a transistor-dependent temperature voltage U_T, whereby U_{out} is in the volt range and U_T in the millivolt range. Since it was assumed above that $\Delta U_{in1} = -\Delta U_{in2}$, the following applies to the differential voltage ΔU_D and according to the above formula for the differential voltage amplification A_D.

$$\Delta U_D = 2 \cdot \Delta U_{in1}$$
$$A_D = \frac{dU_{out1/2}}{dU_D} = \frac{1}{2} \cdot \frac{U_{B+} - U_{out1/2,OP}}{U_T} \tag{4.22}$$

For typical transistors and resistors, A_D is in the range of about 100, so the differential amplification A_D is much higher than the common mode gain A_{cm}. These factors can be further increased by connecting several such differential amplifiers in series, which ultimately leads to the operational amplifiers described below; these play a major role in the A/D interface of embedded systems.

4.7.3   Operational Amplifier

The *operational amplifier*, in the following abbreviated as *OpAmp*, is a multi-level differential amplifier with theoretically infinite gain. As shown above, differential amplifiers offer the possibility of creating large amplifications at very low input currents. However, its outputs have quite high impedance, so they deliver only a limited current. In order to avoid this in practice, they are extended by additional amplifier stages in the operational amplifier. The basic structure is shown in Fig. 4.36.

The differential amplifier (shown on the left) is followed by an additional amplifier stage (in the middle). This increases the amplification factor again, but also provides so-called *frequency compensation*, by means of the integrated capacitor. The idea is that the capacitor, together with the (differential) resistance of the transistor, forms a low-pass filter that suppresses any oscillations that may occur in the circuit. Finally, a so-called *push-pull stage* (on the right in the figure) is added. The npn transistor amplifies the positive half-wave of the signal, the pnp transistor the negative one.

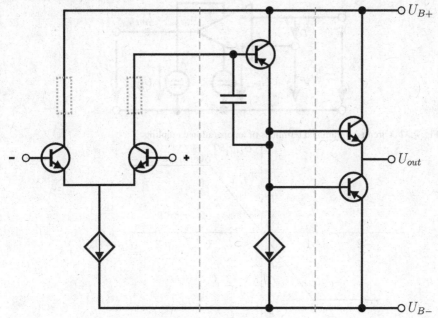

Fig. 4.36 Principle structure of an operational amplifier

This may result in additional amplification, but above all this stage can drive a high output current, which cancels out the disadvantage of the differential amplifier.

The OpAmp got its name from its original field of application, the execution of arithmetic operations in analog computers. It forms the basis of analog operations such as addition, subtraction and integration. Essentially it was used for solving differential equations experimentally. Today, for example, it is the core of A/D converters (see Sect. 7.5) and trigger circuits.

The circuit diagram for the operational amplifier can be seen in Fig. 4.37. It has two inputs, which are marked "+" and "−". They correspond to the two inputs of the differential amplifier. Signals at the positive input are not inverted and amplified with the factor $+A$. On the other hand, the signals at the minus input are inverted (amplification $-A$). The designations "+" and "−" for the inputs thus indicate the sign of the amplification and not the polarity of the applied voltages.

The difference between the input voltages $U_D = U_P - U_N$ is amplified. A minimum difference between these two voltages already causes a maximum output voltage $U_{out} = \pm U_B$, as shown in Fig. 4.38.

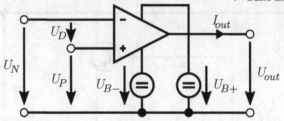

Fig. 4.37 Circuit diagram and terminals of an operational amplifier

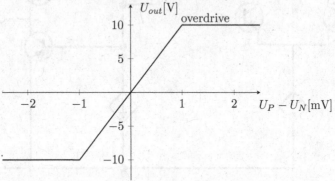

Fig. 4.38 Relation of the input voltages to the output voltage

4.8 Application of Operational Amplifiers

For the analysis of circuits with OpAmps, an infinite gain A can be expected, which in practice ranges from *10,000* to several *100,000*. The common mode gain is smaller by about the factor 10^6 and is therefore negligible.

$$U_{out} = A_0 \cdot (U_P - U_N)$$

The input resistance of the OpAmp can also be regarded as infinitely large. It is typically in the range from 1 to 1000 MΩ. This makes the input currents negligible (typically <0.1 μA) with the result, that control of the OpAmp is practically powerless. Frequencies up to approximately 10 kHz are amplified linearly, i.e. distortion-free.

The most important properties of an OpAmp are summarized in Table 4.3. It should be noted that OpAmps are usually optimized for one of these variables and the other variables are neglected.

Table 4.3 Properties of an operational amplifier

	Ideal	Common
Gain	∞	$10^4 - 2 \cdot 10^5$
Input resistance	∞	$> 10^6 \, \Omega$
Output resistance	0	$< 10 \, \Omega$
Threshold frequency	∞	$> 10 \, \text{kHz}$

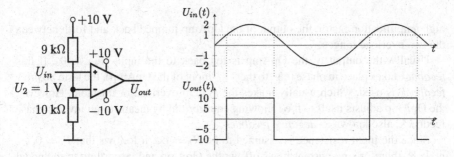

Fig. 4.39 Example circuit for a comparator and excitation with a sine signal

4.8.1 Comparator

The simplest wiring of an OpAmp is that of a threshold switch or comparator. A voltage divider is used to apply a reference voltage U_2 to the negative input. Due to the high input resistance, there is no load on the voltage divider so that the reference voltage results directly from the ratio of the two resistors. This property is used, for example, in sample-and-hold elements (see Sect. 7.4), but also in many analog-to-digital conversion circuits (see Sects. 7.5.1, 7.5.4 and 7.5.6) due to the high gain.

If the actual input voltage U_{in} is greater than the reference voltage U_2, the difference $U_D = U_P - U_N$ is positive and the output of the OpAmp is immediately in the overdrive range ($U_{out} = \min\{U_{B+}, A \cdot (U_{in} - U_2)\} = U_{B+}$).

Figure 4.39 shows a comparator circuit of this sort for $U_B = 10$ V, a reference voltage $U_2 = 1$ V and an approximately sinusoidal input signal U_{in}. As soon as the input signal is greater than 1 V, the output is $U_{out} = 10$ V. If U_{in} falls below the reference voltage $U_2 = 1$ V, the output voltage jumps to $U_{B-} = -10$ V.

4.8.2 Feedback Loop

The mode of operation of the OpAmp is usually determined by its external circuitry. In the previous example, a simple comparator circuit was implemented. Due to the

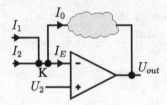

Fig. 4.40 OpAmp with feedback

high amplification factor, the output of the OpAmp jumped back and forth between the two overload ranges.

Usually, the output of the OpAmp is fed back to the input (Fig. 4.40). If this *feedback* takes place in phase (e.g. to the "+"-input of the OpAmp), the term *positive feedback* is used, which usually makes the system overload or unstable. However, the OpAmp adjusts itself to the following equilibrium by means of an inverse phase feedback, also known as *negative feedback*.

Since the input resistance is assumed to be $R_E = \infty$, it follows that $I_E = 0$. At node K, therefore, no current flows off via the OpAmp and, according to the nodal rule it follows, that $I_0 = I_1 + I_2$. Thus, I_0 compensates for the input currents. Since the amplification of an ideal OpAmp is also assumed to be $A = \infty$, U_K is adjusted to U_2: $U_K \rightarrow U_2$. This can be explained as follows

If $U_K < U_2$, the OpAmp would overdrive due to the large gain resulting in $U_{out} \rightarrow +U_B$. Due to the partial feedback of U_{out}, U_K also increases, however, until bis $U_K = U_2$. The same consideration applies to $U_K > U_2$.

With the help of such feedback, inverting and non-inverting amplifiers and more complex computing operators can be realized. Before these are presented, a simple example of a feedback loop will be provided. This is a variant of the comparator introduced above.

4.8.3 *Comparator with Hysteresis (Schmitt Trigger)*

In Fig. 4.41, a circuit for a *Schmitt trigger* is shown. This is named after Otto Schmitt, who invented it in 1934 while still a student. The output alternates between two switching states due to feedback via R_2. If U_D becomes positive, i.e. $U_{in} > U_{out}$, then U_{out} adjusts itself by the positive feedback to $U_{out} = U_B$ and also pulls U_D up (and vice versa). This means that the comparator only switches above or below a limit value that is different from the mean value. The resulting transition zone is called *hysteresis*. This can be determined via the resistances R_1 and R_2.

The Schmitt trigger has a special importance in the area of embedded systems. Since real values do not switch from low to high level immediately, but always take intermediate values, a clean transition can be realized by means of the Schmitt trigger, which is required e.g. for the inverter in Sect. 4.7.1. Trigger circuits are also required

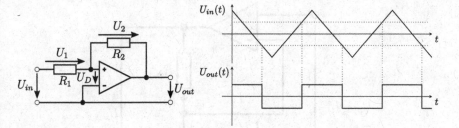

Fig. 4.41 Example circuits for a Schmitt trigger excited by a triangular signal

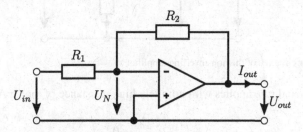

Fig. 4.42 Outer circuitry of the inverting amplifier

whenever a clock or a discrete time division is required, e.g. in oscilloscopes or to generate a PWM-signal (see Sect. 7.7.3).

4.9 Amplifier Circuits

The amplifier circuits used in practice are based on a negative feedback OpAmp. In this case, the output voltage U_{out} is fed back to the negative input. This causes the differential voltage U_D to settle to zero. Depending on the external circuitry, inverting and non-inverting amplifiers can be realized using an OpAmp.

4.9.1 Inverting Amplifier

The basic circuit of the inverting amplifier is shown in Fig. 4.42. When a constant input voltage U_{in} is applied, assuming that $U_{out} = 0$ V, U_N jumps for a short time to the value

$$U_N = \frac{R_2}{R_2 + R_1} \cdot U_{in} \tag{4.23}$$

Due to the high gain A_D and the negative voltage difference at the OpAmp inputs, the voltage U_{out} quickly drops to a negative value. It thus counteracts the positive input voltage via the negative feedback branch until $U_N \approx 0$ V.

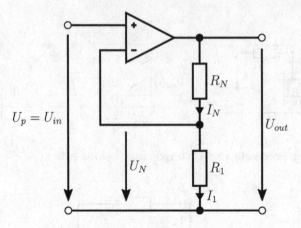

Fig. 4.43 Outer circuitry of the non-inverting amplifier

Then the nodal rule applies with infinitely high resistance of the OpAmp input.

$$\frac{U_{in}}{R_1} + \frac{U_{out}}{R_2} = 0$$

resulting in the overall gain A_U of the circuit

$$A_U = -\left(\frac{U_{out}}{U_{in}}\right) = -\left(\frac{R_2}{R_1}\right) \tag{4.24}$$

Since the negative OpAmp input is practically at ground potential because of $U_N \approx 0$ V, the input resistance R_{in} of the amplifier is

$$R_{in} = \frac{U_{in}}{I_{in}} = R_1 \tag{4.25}$$

U_{out} does not depend on I_{out}, because the OpAmp provides $U_N \approx 0$ V (almost) independently of the output load. For an ideal OpAmp, the output resistance R_{out} is therefore

$$R_{out} = \frac{dU_{out}}{dI_{out}} = 0 \tag{4.26}$$

4.9.2 Non-inverting Amplifier

A disadvantage of the inverting amplifier is its low input resistance. It thus puts load on the output of an upstream circuit and is particularly unsuitable for use as a measuring amplifier for voltage measurements. A non-inverting amplifier (Fig. 4.43) is therefore used to measure voltages without loss.

As with the inverting amplifier, the feedback for the operational amplifier has the following effect

$$U_N \approx U_{in} = U_P$$

Thus, the voltage amplification according to the voltage divider rule applies to the gain A_U.

$$A_U = \frac{U_{out}}{U_{in}} = \frac{R_1 + R_N}{R_1} = 1 + \frac{R_N}{R_1} \qquad (4.27)$$

As with the inverting amplifier, the voltage amplification can also be controlled here by the resistors R_1 and R_N.

The non-inverting amplifier has the very high input resistance of an operational amplifier.

$$R_{in} \Rightarrow \infty \qquad (4.28)$$

U_{out} does not depend on I_{out} because the OpAmp provides $U_N \approx 0$ V (almost) independently of the output load. For an ideal OpAmp, therefore the output resistance settles to

$$R_{out} = \frac{dU_{out}}{dI_{out}} = 0 \qquad (4.29)$$

Due to the two properties that the non-inverting amplifier has a practically infinitely large input resistance and a very large output current can be drawn from it ($R_{out} = 0$), this amplifier is used wherever a source is not to be loaded, but where a lot of current may be required for further processing of the input signal. The circuit can therefore also be used as an impedance converter (isolation amplifier). Often a gain factor $A = 1$ is set simply to separate the input from the output.

4.10 Analog Computation Components

Some circuit variants of the OpAmp are presented below: the analog adder, subtractor and integrator. All three circuits were essential computing elements of analog computers, but are still used today in e.g. analog-to-digital converters, which is why they are presented here in more detail.

4.10.1 Analog Adder

The typical structure of an analog adder is shown in Fig. 4.44. The input currents $I_1, I_2, \ldots$ of the inputs $IN_1, IN_2, \ldots$ are added together in the node K (negative input of the OpAmp), which gives the circuit its name.

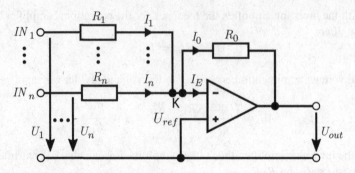

Fig. 4.44 Outer circuitry of the analog adder

Due to the very high input resistance of the OpAmp, its input current I_{in} can be assumed to be zero. This means for the node K, that

$$I_0 = I_1 + \cdots + I_n$$

Furthermore, the differential voltage at the OpAmp input can again be assumed to be zero, so that the voltage at point K results to $U_K = U_{ref} = 0$ V. With the help of Ohm's law, the above node equation can be transformed.

$$I_0 = -\frac{U_{out}}{R_0} = \left(\frac{U_1}{R_1} + \frac{U_2}{R_2} + \cdots + \frac{U_n}{R_n} \right)$$

If this equation is solved for U_{out}, the output voltage is obtained as the weighted sum of the input voltages.

$$U_{out} = - \left(\frac{R_0}{R_1} U_1 + \frac{R_0}{R_2} U_2 + \cdots + \frac{R_0}{R_n} U_n \right)$$

In case that all input resistances are equal to R_0, the weightings are reduced to one and the output voltage results to

$$U_{out} = -(U_1 + U_2 + \cdots + U_n) = - \sum_{k=1}^{n} U_k \tag{4.30}$$

Today, this circuit is used, for example, in the R-2R resistor ladder network for digital/analog conversion (see Sect. 7.7.2).

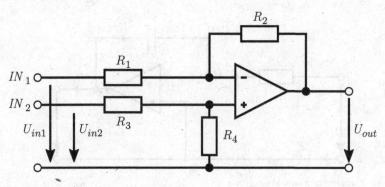

Fig. 4.45 Outer circuitry of the analog subtractor

4.10.2   Analog Subtractor

For analog subtraction, the basic function of the OpAmp, differential amplification, is used. The wiring of the OpAmp is shown in Fig. 4.45. The fact that this circuit actually calculates the difference between the two inputs IN_1 and IN_2 can now be inferred step by step.

   In order to be able to calculate the output voltage of the differential amplifier, the two inputs IN_1 and IN_2 are first considered separately from each other. At first, E_2 is connected to ground. This results in the same output voltage as with the inverting amplifier.

$$U_{out} = -U_{in1} \cdot \frac{R_2}{R_1}$$

If IN_1 is connected to ground, the output voltage is the same as that of the non-inverting amplifier.

$$U_{out} = \frac{R_1 + R_2}{R_1} \cdot \frac{R_4}{R_3 + R_4} \cdot U_{in2}$$

If both inputs are used, the output voltage results to

$$U_{out} = \frac{R_1 + R_2}{R_1} \cdot \frac{R_4}{R_3 + R_4} \cdot U_{in2} - U_{in1} \cdot \frac{R_2}{R_1}$$

If the resistors are selected so that $R_1 = R_3$ and $R_2 = R_4$, the output voltage is

$$U_{out} = \frac{R_2}{R_1} \cdot (U_{in2} - U_{in1})$$

If no amplification is required for the output voltage, all resistors should be of the same size. This simplifies the initial equation to

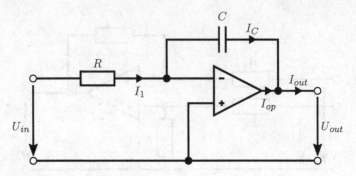

Fig. 4.46 Circuitry of a simple integrator

$$U_{out} = U_{in2} - U_{in1} \tag{4.31}$$

4.10.3 Integrator

The basic circuit of the integrator in Fig. 4.46 is very similar to that of the inverting amplifier. In the feedback branch, the resistor is replaced by a capacitor. The capacitor makes the circuit frequency-dependent. With increasing frequency the impedance of the capacitor decreases and thus the output voltage decreases (low-pass behavior).

As with the inverting amplifier, a virtual ground is assumed at the inverting input of the operational amplifier; as a result, $I_C = I_1$ applies. If a positive voltage U_{in} is applied to the input, the capacitor charges up to the maximum negative output voltage of the operational amplifier with $I_C = \frac{U_{in}}{R}$. The voltage at the capacitor U_C has thus the opposite sign to the output voltage U_{out}. If a negative voltage is applied to the input, the output voltage rises to the maximum output voltage.

If U_{in} is constant, the change of the output voltage ΔU_{out} can be calculated. Let $I_{in} = 0$ A. $I_C = I_1$ and $U_N = 0$ V: $I_1 = U_{in}/R$ apply. The charge on the capacitor results to

$$\Delta Q = I_1 \cdot \Delta t = I_C \cdot \Delta t \tag{4.32}$$

and the voltage across the capacitor to

$$\Delta U_C = \frac{\Delta Q}{C} = \frac{I_C}{C} \cdot \Delta t = \frac{U_{in}}{RC} \Delta t \tag{4.33}$$

Assuming that $U_N = U_{ref}$, the output voltage is calculated to be

$$\Delta U_{out} = -\Delta U_C = -\frac{U_{in}}{RC} \Delta t \tag{4.34}$$

The same derivation can be used for a variable input voltage.

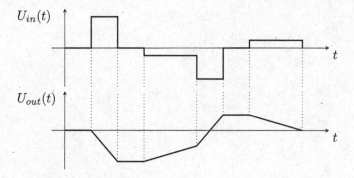

Fig. 4.47 Voltage course at the integrator

If U_{in} is variable, the result is

$$dU_{out} = -\frac{U_{in}}{RC}dt \tag{4.35}$$

$$\Rightarrow U_{out} = -\frac{1}{RC}\int_0^t U_{in}dt + U_{out0}$$

As shown in Fig. 4.47, the output voltage is proportional to the integral over the input voltage.

This circuit is still widely used, for example, in analog-to-digital converters (see Sects. 7.5.4, 7.5.6 and 7.5.7).

References

1. Chen WK (ed) (2004) The electrical engineering handbook. Elsevier, Burlington
2. Svoboda JA, Dorf RC (2013) Introduction to electric circuits. Wiley, Jefferson City
3. Tietze U, Schenk C, Gamm E (2015) Electronic circuits: handbook for design and application, 2nd edn. Springer, Heidelberg

Part II
Systems Theory and Control

Chapter 5
Systems Theory

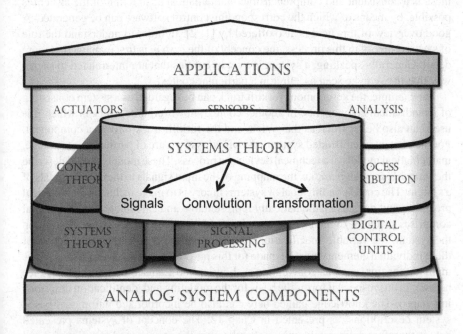

This chapter introduces the theoretical principles required to model and control a technical system, which will be explained in the next chapter. First, the signal is grouped into different classes and the special properties of linear, time-invariant systems (LTI) are explained. Furthermore, convolution is discussed as a way of determining the interaction of two functions, such as signals and transfer functions, in the time domain. Finally, the Fourier and Laplace transformation are used to transform these to the frequency and complex variable domain, respectively, so that even complex systems can be analyzed with less computational effort. This is essential, among other things, for sampling analog signals (see Sect. 7.3.1) and for controller design.

© The Author(s), under exclusive license to Springer Nature Switzerland AG 2021   85
K. Berns et al., *Technical Foundations of Embedded Systems*, Lecture Notes in Electrical
Engineering 732, https://doi.org/10.1007/978-3-030-65157-2_5

5.1 Modeling the Embedded System

Following on from the basic knowledge about electrotechnical circuits and their details provided in the first chapters, the focus is now on the overall system. One of the main tasks of embedded systems is the control of technical systems. The aim is to influence the processes within a system in such a way that the target variable(s) fulfil certain criteria. This can mean, for example, that the system "autonomous forklift truck" is influenced in such a way that the target property "speed" meets the criterion "less than or equal to 10 km/h". This influence is provided by a *controller*. This can be a human "driver", but also an embedded system.

In order to bring about such control, the system to be regulated must first be analyzed and modeled, for which systems theory provides a series of tools. Only with these is a consistent and comprehensible mathematical abstraction of the system is possible, by means of which the corresponding control software can be generated. A good overview of this field is also offered by [1, 2]. In order to understand the role of systems theory in this process, the concept of the *system* is first explained in more detail. Generally speaking, a system is a set of entities that are interrelated in such a way that they can be seen as a unit in a particular context.

For example, the autonomous forklift truck can be regarded as a system consisting of its individual components. In another context, however, the factory in which it is used can also be regarded as the system and the forklift truck only as a component. For the context of embedded systems, *technical systems* are of particular interest, i.e. mathematical models of a technical device or a process. These models should describe the *transmission behavior*, i.e. the mapping of the input signals to the output signals of a system. The corresponding goal of systems theory is to make technical and physical processes and devices mathematically representable and calculable. Since this goal comprises different domains (e.g. the electrical circuits used, the forces acting on the system, the variable specifications of the controller...), a certain abstraction of these individual elements must be made for this purpose. Systems theory uses various methods for this, such as convolution or Fourier and Laplace transformations (see below). In this respect, it forms the basis for the modeling and identification described in Chap. 6. The knowledge gained here can also be incorporated into model-based system descriptions, as presented in Chap. 12. The concept of systems presented here is therefore deliberately universal and covers the entire range of embedded and technical systems. These could be, for example:

- Transmission behavior of a computer network,
- Changing the temperature of a liquid in a heating tank,
- Behavior of a motor vehicle after a change of road surface,
- Reaction of a robot to the approach of a person.

All technical systems to be controlled fall into the class of *dynamic systems*. Another term that is often used in connection with systems theory is that of the *process*. This is not easy to grasp. According to the general ISO definition, a process is

"a set of interrelated or interacting activities that transforms inputs into outputs."

The development of embedded systems focuses in particular on the *technical process*. This describes the part of the overall system to be influenced, for example the kinematics of a robot arm and the positions of the objects in its environment. Here, too, the signals and the transmission behavior of the system are an essential focus of systems theory. It should be noted that the definition of a process used here must not be confused with the software "process", which is used, for example, in Chap. 13 and has a more specialized definition.

5.2 Signals

The components of all embedded systems of interest in this book exchange signals. These can be *analog* or—as is the case with computer communication—*digital*.

The communication structures within embedded systems serve to transmit signals. While in everyday life the term "signal" is generally understood to mean a process to attract attention (e.g. light signal, whistle or wink), a somewhat more limited and precise definition from information technology is used below.

A *signal* is the representation of information by a time-varying physical, in particular electrical value, e.g. current, voltage, field strength. The information is encoded by a parameter for this value, e.g. amplitude, phase, frequency, pulse duration, or a mathematical abstraction thereof.

Signals can be divided into classes according to various criteria. Possible criteria are, for example, whether a signal is stochastic or deterministic, whether it is continuous or discrete, whether it has a finite or infinite duration and whether it has a finite energy or power consumption. The most important signal classes are described below.

5.2.1 Stochastic and Deterministic Signals

Stochastic signals are all non-periodic, fluctuating signals. This class includes all signals occurring in reality, including video, speech and message signals. What they have in common is that they cannot be accurately predicted.

On the other hand, the class of *deterministic signals* includes signals whose course can be unambiguously described by a formula, a table or an algorithm. This makes future values predictable. Although this signal class plays only a minor role in communication practice, it is of great importance for the description of communication systems and for message theory. Many stochastic signals can be approximated sufficiently accurately by deterministic signals, which greatly simplifies the development process. Deterministic signals are subdivided once again by their duration into *transient* or *aperiodic signals* of finite duration, the course of which can be represented over the entire time range (e.g. switch-on process) and into periodic signals of theoretically infinite duration (e.g. sinusoidal or clock signals).

A deterministic signal is completely described by its time function $x(t)$ or its amplitude spectral function $X(\omega)$—amplitude spectrum for short—in the frequency domain (see Sect. 5.4). Both representations are mathematically equivalent. The transition between time and frequency/complex variable domain is made by a mathematical transformation (e.g. Fourier, Z or Laplace transformation, see Sect. 5.4).

5.2.2 Continuous and Discrete Signals

The differentiation of signals into *continuous* and *discrete* signals is essential for the development of embedded systems, especially in communication. The two attributes 'continuous' and 'discrete' concern both the time and the value range of the signal:

- **continuous-time** means that the signal value is defined for each point in time of a (continuous) time interval,
- **discrete-time**, that the signal value is only defined for discrete, mostly equidistant points in time (this includes, for example, signals obtained by polling or discretely sampled signals, see Sect. 7.3),
- **continuous-valued**, that the value range of the signal covers all points of an interval,
- **discrete-valued**, that the value range of the signal contains only discrete function values.

Since the classification of the time and the value domain are independent of each other, there are a total of four possible combinations or signal classes:

Continuous-Time and Continuous-Valued.

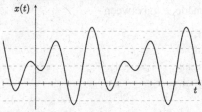

This is referred to as an *analog* signal. This applies, for example, to the output voltage of a charging capacitor (see Sect. 3.4.1.1).

Discrete-Time and Continuous-Valued.

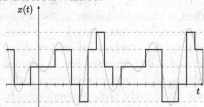

This is referred to as a *sampling signal*. (The signal is described by the dots, not by the dashed lines.) It is created, for example, as an intermediate step in the digitization of analog signals (see Sect. 7.3).

Continuous-Time and Discrete-Valued.

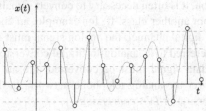

This is referred to as an *amplitude-quantized* signal. This appears, for example, at the end of a digital/analog conversion (see Sect. 7.7).

Discrete-Time and Discrete-Valued.

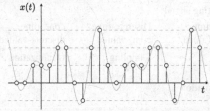

This is referred to as a *digital* signal. (The signal is described by the dots, not by the dashed lines). These signals do not occur in reality, but the signals used in digital technology are approximated as such.

In colloquial language, digital signals and binary signals are often used interchangeably, but although binary signals are the most commonly used digital signals, there are also other forms. Depending on the size M of the value set of a digital signal, a distinction is made, e.g. between

- $M = 2$: binary signal,
- $M = 3$: ternary signal,
- $M = 4$: quaternary signal,
- $M = 8$: octernary signal,
- ...

Such signals are mainly found in transmission of messages, but they also appear as intermediate stages, for example in the conversion of analogue into binary, digital signals.

In signal transmission, it is often necessary to convert signals from one of the four classes into signals from another class. If, for example, an analog speech signal is to be transmitted via a digital channel, the analog signal must be transformed into a digital signal. This is referred to as analog-to-digital conversion or *A/D conversion*. Conversely, a digital data signal must be converted into an analog signal before it can be transmitted over the analog telephone network. In this case we speak of digital-to-analog conversion or *D/A conversion*.

The possible conversion methods of the signal classes are shown in Table 5.1. There is no separate term for the back colored fields, since these transformations do not occur in practice.

Due to the physical properties, only continuous-time signals can be transmitted via real transmission media. If one further considers that the real transmission channels also show a low-pass effect, a transmitted signal is also always continuous in value. It follows from these considerations that a *physically transmitted signal is always an analog signal*. Therefore, the problems of analog signal transmission cannot be completely ignored in digital data transmission. In addition, analog transmission systems must therefore also be taken further into account in the development of actual digital embedded systems. Conversely, analog signals must be digitized before they are processed in these systems.

Table 5.1 Possible transformations between classes of signals

Original signal	Resulting signal			
	cont.-time, cont.-valued	Discrete-time, cont.-valued	cont.-time, discrete-valued	Discrete-time, discrete-valued
cont.-time, cont.-valued	–	Sampling	Quantization	A/D conversion
Discrete-time, cont.-valued	Interpolation	–		Quantization
cont.-time, discrete-valued	Smoothing		–	Sampling
Discrete-time, discrete-valued	D/A conversion	Quantization	Interpolation	–

5.2.3 Energy and Power Signals

Every physical transmission of information or signals requires energy or power. In addition to the division of signals into continuous-time/-valued or discrete signals, a classification into *energy signals* and *power signals* is also important for data transmission. This plays a particularly important role in subsequent modeling. The following example is intended to explain these terms.

A very simple transmission line model should be used here to calculate the energy or the average power during signal transmission via a physical medium. It is only considered how the electrical energy or power is calculated at an ohmic resistance. A simple line model is shown in Fig. 5.1.

The energy E provided in a time interval t_1 to t_2 at the resistor R is

$$E = \int_{t_1}^{t_2} U(t)I(t)dt = \frac{1}{R}\int_{t_1}^{t_2} U^2(t)dt = R\int_{t_1}^{t_2} I^2(t)dt \qquad (5.1)$$

The electrical energy delivered at the resistor R is therefore proportional to the integral across the square of a time function, in this case either the voltage $U(t)$ or the current $I(t)$ that varies with time. Accordingly, the average power P output at the resistor R is defined as

$$P = \frac{1}{2t}\int_{t_1}^{t_2} U(t)I(t)dt = \frac{1}{2t}\frac{1}{R}\int_{t_1}^{t_2} U^2(t)dt = \frac{1}{2t}R\int_{t_1}^{t_2} I^2(t)dt \qquad (5.2)$$

The two formulas can now be used to define what an energy or power signal is. The term energy signal is used when the energy delivered to the resistor is finite over an infinite time interval.

$$0 < E = \int_{-\infty}^{\infty} x^2(t)dt < \infty \qquad (5.3)$$

Accordingly, a signal is a power signal if the average power over the entire time range is finite.

$$0 < P = \frac{1}{2t}\lim_{t\to\infty}\int_{-t}^{t} x^2(t)dt < \infty \quad \text{with } x(t): I(t) \text{ or } U(t) \qquad (5.4)$$

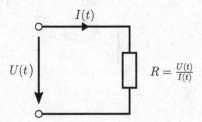

Fig. 5.1 Energy and power at a resistor

It is easy to show that a power signal cannot be an energy signal, because in this case the energy emitted in infinite time intervals would be infinite. Conversely, no energy signal is a power signal, since the average power then becomes zero in the infinite time interval.

The functions shown below are some examples of energy and power signals. In principle, direct signals and periodic signals fall into the class of power signals and transient/aperiodic signals (deterministic signals of finite duration) are classified as energy signals.

Energy Signals

Square pulse

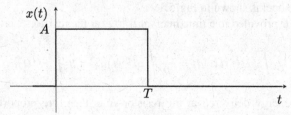

$$E = A^2 T$$

$$x(t) = \begin{cases} A, & 0 \le t \le T \\ 0, & else \end{cases}$$

Exponential function

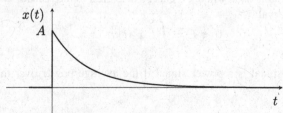

$$E = \int_0^\infty e^{-2\alpha t} dt = \frac{-1}{2\alpha} e^{-2\alpha t} \big|_0^\infty = \frac{1}{2\alpha}$$

$$x(t) = \begin{cases} e^{-at}, & t \ge 0 \\ 0, & else \end{cases}$$

Power Signals
Direct signal

$$P = A^2$$

$$x(t) = A, \quad \forall t$$

Sine function

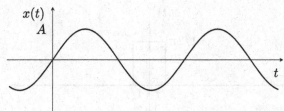

$$P = \frac{1}{2}A^2$$

$$x(t) = A \cdot sin(2\pi f_0 t)$$

Discrete-time signals, whether continuous-valued or discrete-valued (the latter are digital signals), are neither energy nor power signals, since both the energy delivered in the infinite time interval and the power are zero. However, since physical signal transmission is only possible by using energy, digital signals cannot be transmitted via real physical channels and must therefore be approximated by superimposing continuous-valued signals.

Discrete-time signal

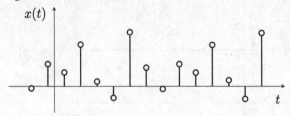

$$E = P = 0$$

An important signal of finite duration for telecommunications is the *Dirac delta function* $\delta(t)$. This signal is a square wave pulse of arbitrarily short duration $2/\tau$ with an amplitude of $0.5/\tau$.

Dirac delta function

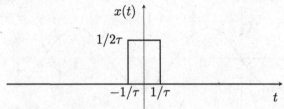

$$P = \lim_{\tau \to \infty} \frac{1}{2\tau} \left[\frac{1}{4}\tau^2 \frac{2}{\tau} \right] = \frac{1}{4}$$

$$x(t) = \delta(t) = \begin{cases} 0, & t \le -\frac{1}{\tau}, \frac{1}{\tau} \le t \\ 0.5\tau, & -\frac{1}{\tau} \le t \le \frac{1}{\tau} \end{cases}$$

The time interval $[-1/\tau, 1/\tau]$ must be regarded as arbitrarily small, i.e. $\tau \to \infty$. Although the Dirac impulse is a finite signal, it belongs to the class of power signals because (by definition) the average power is finite. The Dirac impulse is important for the description and analysis of transmission systems, even if it cannot be generated physically.

5.2.4 Description of Signals in the Time Domain

Signals that allow a particularly simple mathematical description and can be easily generated technically are called *elementary signals*. Important elementary signals

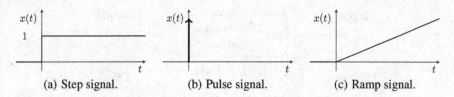

(a) Step signal. (b) Pulse signal. (c) Ramp signal.

Fig. 5.2 Important elementary signals

for system investigations are the step signal ($\sigma(t)$), the ideal pulse signal (Dirac delta function, $\delta(t)$), the ramp signal and the harmonic oscillation.

A *step signal* that occurs at time $t = 0$ (represented in Fig. 5.2a) is defined as follows.

$$x(t) = \sigma(t) = \begin{cases} 0 \ \forall t < 0 \\ 1 \ \forall t \geq 0 \end{cases} \tag{5.5}$$

The Dirac impulse (see Fig. 5.2b) is defined by

$$\int_{-\infty}^{\infty} \delta(t)dt = 1 \tag{5.6}$$

It is a good approximation of a short-term influence on a system.

The *ramp signal* (see Fig. 5.2c) can be defined as the integral of the jump function.

$$x(t) = \int_{-\infty}^{t} \sigma(\tau)d\tau = \begin{cases} 0 \ \forall t < 0 \\ t \ \forall t \geq 0 \end{cases}$$

Finally, the harmonic oscillation is represented by a sinusoidal signal (see Fig. 5.3). Sinusoidal signals of different frequencies are often used to excite systems. The periodic elementary signal is described by

$$x(t) = x_0 \sin(\omega_0 t)$$

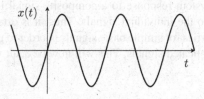

Fig. 5.3 Harmonic oscillation

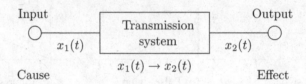

Fig. 5.4 Transmission system

5.2.5 Transmission Systems

Systems theory defines a (transmission) system as a reality oriented mathematical model for the description of the transmission behavior of a complex arrangement. The model is a mathematically unambiguous assignment of an input signal to an output signal, as shown in Fig. 5.4. The assignment is usually referred to as transformation.

For a given real system, the relationship between system excitation (input signal, cause), the system itself and the effect (system output variable) is searched. While systems theory is generally interested in systems of any complexity, only passive signal transmission media such as cables and radio, and physical systems are considered in the following. These fall into the class of linear, time-invariant systems.

A typical example of such an (embedded) system is the model of a double-wire line, which is very common in practice. Examples in message exchange would be twisted pair lines (see Sect. 11.4.2) in telecommunications and in local area networks (LANs). A simple model of such a double-wire line is shown in Fig. 5.5. This compact model describes a line quite accurately as long as the line is short compared to the wavelength of the transmitted signal.

Some simplifications are often assumed for the analysis of the transmission behavior of the double-wire line. For example, the internal resistance of the signal source R_i is assumed to be zero and the load resistance R_l that models the signal sink is set to infinity. A jump signal or a square wave signal is usually used as the input signal.

Depending on the frequency of the signal and the design of the cable, it may also be possible to dispense with one or other of the components for modeling the cable (e.g. the coil). A great advantage in the system analysis of this two-wire line is its linear transmission behavior, which can also be found in most physical/technical systems. This means that the system response to a composite signal is the composition of the system responses to the individual signals. Thus, it is often possible simply to study the system properties for simple basic signals in order to gain knowledge of the behavior of more complicated signals. This will now be dealt with more precisely.

5.2.6 Linear, Time-Invariant Systems

All electrical systems consisting of linear, passive components such as resistors, capacitors and coils are linear and time-invariant, e.g. the low-pass filter (see

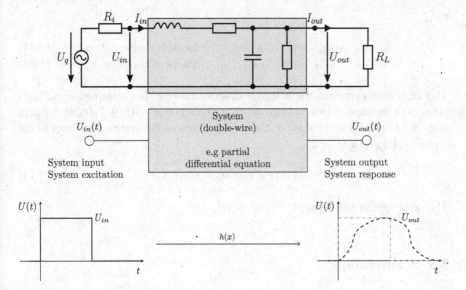

Fig. 5.5 Model of a double-wire line

Sect. 3.5.1) or the harmonic oscillator (see Sect. 6.2.2.1). Many other systems, such as mechanical systems consisting of springs, ground and dampers, also fall into this category. These systems represent a large proportion of real systems. But also non-linear systems can be approximated by means of linearization around the operating point for many applications as linear systems. Linear, time-invariant systems (LTI systems) therefore form an important basis for modeling. The following two definitions describe what linear and time-invariant mean.

A (transmission) system is called *time-invariant*, if for every fixed value t_0 and any signal $x_1(t)$ the following applies.

$$\text{if } x_1(t) \rightarrow x_2(t) \tag{5.7}$$
$$\text{then } x_1(t + t_0) \rightarrow x_2(t + t_0) \tag{5.8}$$

In principle, this definition says nothing other than that a time-invariant transmission system behaves the same at all times. If the system response to an exciter signal x_1 equals x_2 at time t, then the system response to x_1 at any other time $t + t_0$ is also x_2.

A (transmission) system is called *linear* if for any constant a and any signal $u_1(t)$, $v_1(t)$ the following applies.

$$\text{if } u_1(t) \rightarrow u_2(t) = G\{u_1(t)\} \tag{5.9}$$
$$\text{and } v_1(t) \rightarrow v_2(t) = G\{v_1(t)\} \tag{5.10}$$

then

$$G\{u_1(t) + v_1(t)\} = G\{u_1(t)\} + G\{v_1(t)\} \quad \text{(Principle of superposition) and} \quad (5.11)$$
$$G\{a \cdot u_1(t)\} = a \cdot G\{u_1(t)\} \quad \text{(Principle of proportionality)} \quad (5.12)$$

This definition expresses the behavior described above. If a complex signal $x(t)$ consists of the superposition of two simple signals $x(t) = u_1(t) + v_1(t)$, the system response G to $x(t)$ is the same as the superposition of the system responses to the signals $u_1(t)$ and $v_1(t)$, i.e.

$$Gx(t) = Gu_1(t) + Gv_1(t) \tag{5.13}$$

The same applies to a signal $x(t) = a \cdot u_1(t)$.

5.3 Convolution

The elementary signals described in Sect. 5.2.4 are very well suited for determining the transfer function of a system, i.e. the mathematical relationship between the input signals of the system and the resulting output signals. However, in practice it is often the case that a system with a known transfer function or impulse response is present and you want to calculate what the system output looks like for a certain input signal. This procedure is used, for example, for functional testing of components and also plays an important role in the design of controllers, since it is the only way to calculate the behavior of the controller and the controlled system (see Chap. 6) connected in series.

In order to calculate the output signal of a system for any input signal, the *convolution integral* is therefore introduced.

$$y(t) = \int\limits_{-\infty}^{\infty} x(\tau) \cdot g(t - \tau)d\tau = x(t) * g(t) \quad \text{read as: x convolved (with) g} \quad (5.14)$$

with y(t): Output signal, x(t): Input signal, g(t): Impulse response of the system, τ: Shift on the time axis.

In order to derive the convolution integral, the system response $g^{(n)}(t)$ to a single square pulse $r^{(n)}(t)$ with the area 1 is first considered.

$$r^{(n)}(t) = \begin{cases} \frac{n}{T}, & 0 \le t < \frac{T}{n} \\ 0 & \text{else} \end{cases} \tag{5.15}$$

As the definition of the square pulse $r^{(n)}(t)$ shows, the pulse duration becomes smaller and smaller as n increases, while the area of the square-wave remains constant.

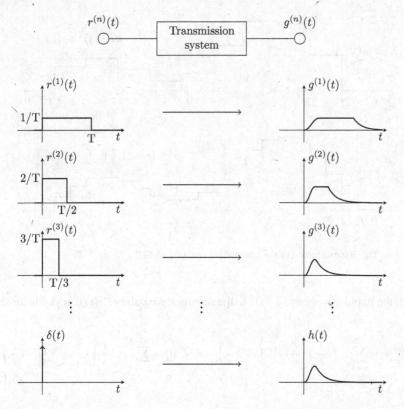

Fig. 5.6 Transition of the square function to the Dirac impulse including the respective system responses

For $n \to \infty$ the square pulse is accordingly converted into the Dirac impulse (see Fig. 5.6).

The response $x_2(t)$ to a general input signal $x_1(t)$ is now calculated. To do this, the signal $x_1(t)$ is approximated by means of square-wave pulses, the pulse widths of which become narrower and narrower until it transitions from the square pulse to the Dirac impulse, as follows.

It is assumed that $x_1(t)$ has a length (duration) of $L \cdot T$, $x_1(t)$ is approximated by the sum of shifted, weighted square pulses $r^{(n)}(t)$ and $\tilde{x}_1^{(n)}(t)$ designates the signal approximation of $x_1(t)$.

The result of the above formula is that $r^{(n)}(t - k\frac{T}{n})$ is only not equal to zero in the interval $k\frac{T}{n} \le t < (k+1)\frac{T}{n}$. Accordingly, the following sum of individual squares approximates the signal $x_1(t)$.

$$\tilde{x}_1^{(n)}(t) = \sum_{k=1}^{nL} x_1\left(k\frac{T}{n}\right) \cdot \frac{T}{n} \cdot r^{(n)}\left(t - k\frac{T}{n}\right) \tag{5.16}$$

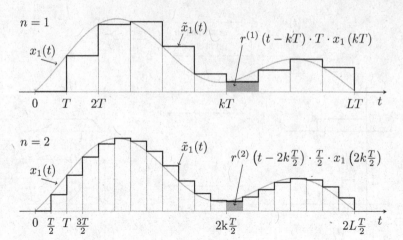

Fig. 5.7 The first to addends of the convolution sum (Eq. (5.17))

With the impulse response $g^{(n)}(t)$ to the square-wave pulse $r^{(n)}(t)$ this yields for any n

$$\tilde{x}_1^{(n)}(t) = \sum_{k=1}^{nL} x_1\left(k\frac{T}{n}\right) \cdot \frac{T}{n} \cdot r^{(n)}\left(t - k\frac{T}{n}\right) \longrightarrow \tilde{x}_2^{(n)}(t) = \sum_{k=1}^{nL} x_1\left(k\frac{T}{n}\right) \cdot \frac{T}{n} \cdot g^{(n)}\left(t - k\frac{T}{n}\right)$$

(5.17)

Now, n is allowed to become arbitrarily large, i.e. the pulse width of the square-wave pulses becomes arbitrarily narrow (Fig. 5.7 shows the first two addends graphically). For $n \to \infty$ the square wave pulse becomes the Dirac impulse. In particular

- $\frac{T}{n} = d\tau; k\frac{T}{n} = \tau$,
- $r^{(n)}(t) \to \delta(t)$ (Dirac-pulse),
- $g^{(n)}(t) \to h(t)$ (system response).

With $n \to \infty$ the signal approximations $\tilde{x}_1^{(n)}(t)$ and $\tilde{x}_2^{(n)}(t)$ converge with the signals $x_1(t)$ and $x_2(t)$, since the area of the Dirac impulse $\delta(t - \tau)$ shifted by τ is equal to one.

$$\lim_{n\to\infty} \tilde{x}_1^{(n)}(t) = \int_0^{LT} x_1(\tau) \cdot \delta(t - \tau)d\tau = x_1(t) \cdot \int_0^{LT} \delta(t - \tau)d\tau = x_1(t) \quad (5.18)$$

$$\lim_{n\to\infty} \tilde{x}_2^{(n)}(t) = \int_0^{LT} x_1(\tau) \cdot h(t - \tau)d\tau = x_2(t) \quad (5.19)$$

For signals that are not time-limited, the convolution results to

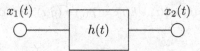

Fig. 5.8 Graphical representation of a general transmission system

$$x_2(t) = \int\limits_{-\infty}^{\infty} x_1(\tau) \cdot h(t-\tau)d\tau = x_1(t) * h(t) \quad (x_1 \text{ convolved with h}) \quad (5.20)$$

With the help of this convolution integral, it is now possible to determine the system response $x_2(t)$ to the exciter signal $x_1(t)$ when the system response to the Dirac impulse $\delta(t) \to h(t)$ is known (see Fig. 5.8).

As can be seen in Eq. (5.20), the Dirac impulse as an input signal $x_1(t) = \delta(t)$ results in the impulse response as a *characteristic system response* to

$$x_2(t) = \int\limits_{-\infty}^{\infty} \delta(\tau) \cdot h(t-\tau)d\tau = h(t) \quad \text{or} \quad g(t) \quad (5.21)$$

This characteristic system response (also called *impulse response* or *pulse response*) plays an important role in system identification, since it is already possible to make estimates of the general system behavior on the basis of the reaction of the system via an excitation with a Dirac impulse.

When using the convolution integral, the calculation rules shown in the following list apply.

Commutativity

$$f(t) * g(t) = g(t) * f(t)$$

Associativity

$$f(t) * g(t) * h(t) = f(t) * (g(t) * h(t))$$

Distributivity

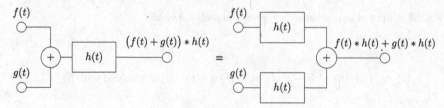

$$(f(t) + g(t)) * h(t) = f(t) * h(t) + g(t) * h(t)$$

5.3.1 *Graphical Representation of the Convolution*

In addition to the purely arithmetic folding solution presented above, there is also a graphical representation and solution method. This is particularly useful for simple or sectionally defined transfer functions and signals, since the result can often be seen directly without a complex calculation, or the actual calculation can be split into simpler partial calculations.

The following procedure is applied in this context. First the impulse response is mirrored at the origin, then the signals are "shifted" into each other, as shown in Fig. 5.9. The (partial) convolution integrals are set up and calculated.

This graphical approach will be illustrated by two examples. The excitation signal selected is a square-wave pulse $x_1(t)$, which is typical for digital data transmission. The impulse response $h(t)$ from Fig. 5.10 is also a square-wave signal here. The system response $x_2(t)$ to the excitation with $x_1(t)$ is being calculated. To find this, the functions $x_1(t)$ and $h(t)$ have to be convolved.

For this purpose, the input signal is first mirrored at the origin, which in the case of a square-wave signal simply corresponds to a shift to the other side of the axis. This signal is then shifted from left to right via the impulse response function (see Fig. 5.10c). Wherever the two signals overlap (in the figure, for example, from time 0 to t), the appropriate convolution integral is calculated (here: $x_2(t) = \int_0^t h(\tau) \cdot x_1(t - \tau)d\tau = \hat{X}t$) and its validity limits determined, i.e. the points in time between which this integral is valid. This results in different sections for which the respective convolution integral can now be calculated.

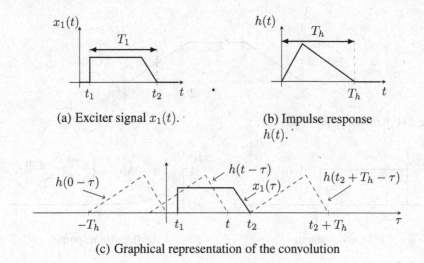

(a) Exciter signal $x_1(t)$.

(b) Impulse response $h(t)$.

(c) Graphical representation of the convolution

Fig. 5.9 Example of the graphical representation of the convolution

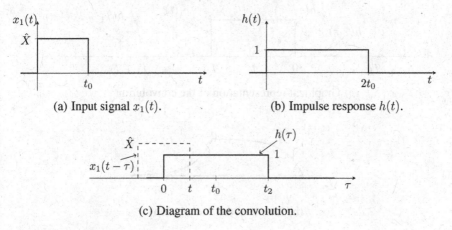

(a) Input signal $x_1(t)$.

(b) Impulse response $h(t)$.

(c) Diagram of the convolution.

Fig. 5.10 Convolution of $x_1(t)$ and $h(t)$

$$x_2(t) = \int_{\tau=0}^{t\le t_0} h(\tau) \cdot x_1(t-\tau)d\tau = \begin{cases} 0 & t \le 0 \\ \hat{X}t & 0 \le t \le t_0 \\ \hat{X}t_0 & t_0 \le t \le 2t_0 \\ \hat{X}(3t_0 - t) & 2t_0 \le t \le 3t_0 \\ 0 & 3t_0 \le t \end{cases} \quad (5.22)$$

Therefore, the curve of the output signal shown in Fig. 5.11 emerges.

A further example is shown in Fig. 5.12. Another square-wave pulse of duration T_1 is given as an excitation signal $x_1(t)$ for a system with an exponential function

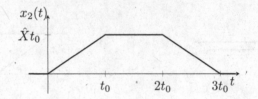

Fig. 5.11 Resulting output signal

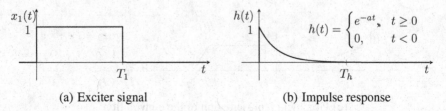

(a) Exciter signal   (b) Impulse response

Fig. 5.12 Exciter signal and impulse response—example

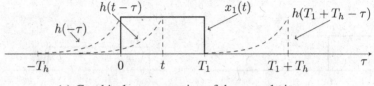

(a) Graphical representation of the convolution

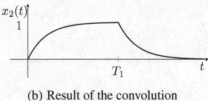

(b) Result of the convolution

Fig. 5.13 Graphical representation of a convolution—example

$h(t)$ as its impulse response, as is typical for passive lines, for example (cf. also the switching process at the low-pass filter in Sect. 3.5.1.2).

To calculate the convolution integral with the integration variable τ, the two functions $x_1(t)$ and $h(t)$ are applied again to the τ axis, this time mirroring of $h(t)$ on the axis takes place according to the commutativity law. For $t = 0$ the result is $h(-\tau)$. The impulse response appears mirrored or convolved (hence the name *convolution*). For $t > 0$ the time-inverse impulse response is shifted to the right, as shown in Fig. 5.13a. For each position, the integral of the product $s_1(\tau) \cdot h(t - \tau)$ is created. This results in the output signal $x_2(t)$ shown in Fig. 5.13b.

5.3.2 Discrete-Time Convolution

Since today most controls are implemented on a digital system, such as a micropro-cessor, the signals to be convolved are usually time-discrete, as shown in Sect. 5.2.2, such as the signal represented in Fig. 5.14. If this is the case, certain special features arise in the calculation, which also play an important role in embedded systems because of the lower computing effort involved. Instead of the integral, only sums are formed and the time dependence can be represented by counters.

The length L of a sequence of time-discrete signals $\{x(n)\}_a^b = \{x(a), \ldots, x(b)\}$ equals

$$L = b - a \tag{5.23}$$

So a signal has $L + 1$ sequential elements.

Discrete-time sequences in the time domain are

$$\{x_1(n)\}, \{h(n)\}, \{x_2(n)\}$$

Corresponding discrete-time sequences in the frequency domain are

$$\{X_1(k)\}, \{H(k)\}, \{X_2(k)\}$$

The discrete nature of the signals therefore simplifies convolution in the time domain to

$$\{x_2(n)\} = \left\{ \sum_{\nu=-\infty}^{\infty} x_1(\nu) \cdot h(n - \nu) \right\} = \{x_1(n)\} * \{h(n)\} \tag{5.24}$$

with the impulse response $\{h(n)\}$ = system response to the unit impulse at the input $\{x_1(n)\} = \{\delta(n)\}$ with

$$\{\delta(n)\} = \begin{cases} 1 & \text{for } n = 0 \\ 0 & \text{else} \end{cases}$$

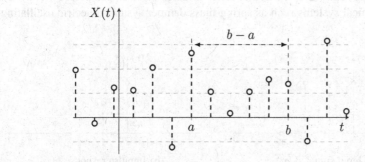

Fig. 5.14 Discrete-time signal

The following is an example of discrete convolution. Let $x_1(n)$ and $h(n)$ be the given input signals corresponding to Fig. 5.15. The system response $x_2(n)$ to excitation with $x_1(n)$ is being calculated by means of convolution operations.

Calculating the convolution sum

$$x_2(n) = x_1(n) * h(n)$$

From Eq. (5.24)

$$x_2(n) = \sum_{\nu=0}^{n} x_1(\nu) \cdot h(n - \nu)$$

Since the signal $h(n)$ has only 5 values, this can be calculated in single steps.

$$
\begin{aligned}
x_2(0) &= x_1(0) \cdot h(0) = 1 \\
x_2(1) &= x_1(1) \cdot h(0) + x_1(0) \cdot h(1) = 1.6 \\
x_2(2) &= x_1(2) \cdot h(0) + x_1(1) \cdot h(1) + x_1(0) \cdot h(2) = 2 \\
x_2(3) &= x_1(2) \cdot h(1) + x_1(1) \cdot h(2) + x_1(0) \cdot h(3) = 1.2 \\
x_2(4) &= x_1(1) \cdot h(3) + x_1(2) \cdot h(2) = 0.6 \\
x_2(5) &= x_1(2) \cdot h(3) = 0.2
\end{aligned}
$$

The output signal is shown in Fig. 5.16.

5.4 Modeling Dynamic Systems

In order to describe and calculate the properties of a technical system, it must first be converted into a mathematical model. Since the temporal behavior of the system is usually relevant in control engineering, a dynamic model is required here in which this behavior is described. This time dependence is often not directly proportional, but only occurs through the derivation of state variables (such as dampers: $F_d = d \cdot \dot{x}$). Simple technical systems such as spring-mass damper systems, electric oscillating

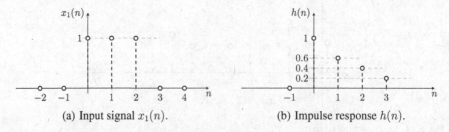

(a) Input signal $x_1(n)$.   (b) Impulse response $h(n)$.

Fig. 5.15 Signal curve of $x_1(n)$ and $h(n)$

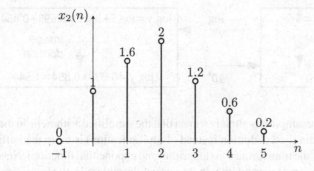

Fig. 5.16 System response $x_2(n)$, obtained by graphical convolution

circuits or double-wire lines are described by linear differential equations with fixed coefficients.

$$b_n^{(n)} x_{out}(t) + \cdots + b_1 \dot{x}_{out}(t) + b_0 x_{out}(t) = a_0 x_{in}(t) + a_1 \dot{x}_{in}(t) + \cdots + a_n^{(m)} x_{in}(t)$$

The solution of such differential equations is generally very complex. In order to simplify them and to analyze more complex transmission systems, a transformation from the time domain to the frequency domain is performed. For this purpose, transformations such as the *Fourier* and *Laplace* transformations presented below are performed. Further transformations and a deeper insight into the matter are offered, for example, by [1, 3].

With the transition from the time to the frequency domain, complex operations of differentiation and integration (e.g. the convolution described above) change into simpler algebraic operations (e.g. multiplication and division) with complex variables. Complex systems of differential equations then become relatively easy to solve, linear systems of equations (cf. the example in Sect. 6.2.2.1). After analysis of the transmission system in the frequency domain, however, the result must be transformed back into the time domain. Even if the two transformations at the beginning and end of the system analysis require a certain effort, the entire system analysis in the frequency domain is usually simpler than directly in the time domain.

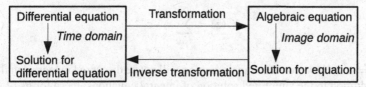

The logarithm serves to clarify the underlying idea of such a transformation. With its help, a multiplication can be mapped to a simpler addition in the complex variable domain. This property was exploited until the 1980s when working with a slide rule, before it was replaced by today's calculator.

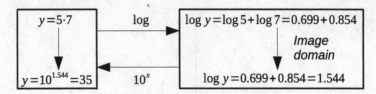

In this simple example it is already shown that the simplified arithmetic in the complex variable domain (here: addition instead of multiplication) is more than offset by the complexity of the transformation (logarithm and exponential function). Nevertheless, the procedure can be worthwhile. In the case of the slide rule, the transformation was printed once on the slide rule by a logarithmic scale and did not have to be carried out by the user.

So-called integral transformations are therefore used to model technical systems. If $x(t)$ is any function, the integral transformation is calculated as follows.

$$X(\rho) = \int_{t_1}^{t_2} x(t) \cdot K(\rho, t) dt$$

The term $K(\rho, t)$ within the integral is called the core. This differs in the different transformations. In the Fourier and Laplace transformation, the core of the transformation integral looks like this:

Fourier transformation: $K(\omega, t) = e^{-j\omega t}$ $\omega = 2\pi f$ (Angular frequency)

$$\mathcal{F}\{x(t)\} = X(\omega) = \int_{-\infty}^{\infty} x(t) \cdot e^{-j\omega t} dt$$

Laplace transformation: $K(s, t) = e^{-st}$ $s = \alpha + j\omega$

$$\mathcal{L}\{x(t)\} = X(s) = \int_{0}^{\infty} x(t) \cdot e^{-st} dt$$

$s = \alpha + j\omega$ is a complex variable and is often referred to as *complex frequency*.

Another transformation is the Z transformation. This is a special case of the Laplace transformation for time-discrete signals. Its transformation function is

$$X_z(z) = \sum_{k=0}^{\infty} x_k \cdot z^{-k} \tag{5.25}$$

with $z = e^{Ts}$. However, this transformation will not be discussed in more detail here.

In addition to the simplified solution of integral equations, these forms of transformation have some other advantages which will be discussed in more detail below, as will their implementation.

5.4.1 Fourier Series and Transformation

This section now shows how any analog signal can be described by series of elementary functions. Series decomposition is of interest because the passive systems used in practice for transmission have a linear behavior. If the decomposition of a complex signal into a superposition of elementary functions is known, this means that only the transfer behavior for the (simple) elementary functions must be determined in order to calculate the transfer behavior for the complex signal too.

If $x(t)$ is any continuous-time signal, $x(t)$ can be approximated by superimposing the elementary functions $k(t)$, whereby the approximation error becomes smaller the more elementary functions are used for approximation.

$$\tilde{x}(t) = \sum_k c_k \cdot \Phi_k(t) \quad \text{with} \quad \tilde{x}(t) : \text{Approximation of } x(t)$$

$$\Phi_k(t) : \text{Elementary functions, } k = 0, 1, 2, \ldots$$

$$c_k : \text{constant coefficients}$$

Two typical elementary functions are the rectangular function and the sine function. The latter leads to the description of periodic signals by Fourier series or to the description of more general aperiodic signals (finite duration) that usually occur in practice by the Fourier transformation. The Fourier series describes a periodic signal $s(t)$ either

- as a sum of sine and cosine oscillations of different frequencies

$$x(t) = \frac{A_0}{2} + \sum_{n=1}^{\infty} A_n \cos(n\omega_0 t) + \sum_{n=1}^{\infty} B_n \sin(n\omega_0 t) \quad \text{with} \quad \omega_0 = 2\pi f_0 \quad (5.26)$$

- or as a sum of cosine functions of different frequencies and phase positions

$$x(t) = \frac{A_0}{2} + \sum_{n=1}^{\infty} C_n \cos(n\omega_0 t - \varphi_n) \text{ with } C_n = \sqrt{A_n^2 + B_n^2}$$

$$\varphi_n = \arctan\left(\frac{B_n}{A_n}\right)$$

To determine the coefficients, Eq. (5.26) is multiplied with $\cos(m\omega_0 t)$ and integrated by $[-\pi, \pi]$.

$$\int_{-\pi}^{\pi} x(t) \cos(m\omega_0 t) dt = \int_{-\pi}^{\pi} \frac{A_0}{2} \cos(m\omega_0 t) dt + \sum_{n=1}^{\infty} \int_{-\pi}^{\pi} (A_n \cos(n\omega_0 t) + B_n \sin(n\omega_0 t)) \cos(m\omega_0 t) dt$$

With $m = 0$ and $n \neq 0$, it follows from the orthogonality relationships of the trigonometric functions

$$\int\limits_{-\pi}^{\pi} x(t)dt = \int\limits_{-\pi}^{\pi} \frac{A_0}{2}dt = A_0\pi$$

and so $A_0 = \frac{1}{\pi}\int\limits_{-\pi}^{\pi} x(t)dt$. Thus, A_0 describes the direct component, i.e. the part of the signal that remains unchanged over time.

For $m > 0$ and $n = m$, it thus follows

$$A_n = \frac{1}{\pi}\int\limits_{-\pi}^{\pi} x(t)cos(n\omega t)dt$$

and

$$B_n = \frac{1}{\pi}\int\limits_{-\pi}^{\pi} x(t)sin(n\omega t)dt$$

Here, ω_0 is the fundamental or first harmonic oscillation with $\omega_0 = 2\pi f_0 = \frac{2\pi}{T}$ and $n\omega_0$ is the (n-1)th or respectively the nth harmonic. A limitation of this representation is that $n\omega_0 \in \mathcal{N}$. This is circumvented by another representation. From Euler's formula $e^{jx} = cos(x) + jsin(x)$, it follows

$$A_n cos(n\omega_0 t) + B_n sin(n\omega_0 t) = \frac{1}{2}(A_n - jB_n) \cdot e^{jn\omega_0 t} + \frac{1}{2}(A_n + jB_n) \cdot e^{-jn\omega_0 t}$$

By combination of $\frac{1}{2}(A_n - jB_n)$ to the new coefficient C_n and $\frac{1}{2}(A_n + jB_n)$ to C_{-n}, the result is

$$C_n e^{jn\omega_0 t} + C_{-n}e^{-jn\omega_0 t} = A_n cos(n\omega_0 t) + B_n sin(n\omega_0 t)$$

or in the sum

$$\sum_{n=-\infty}^{\infty} C_n e^{jn\omega_0 t} = \frac{A_0}{2} + \sum_{n=1}^{\infty}(A_n cos(n\omega_0 t) + B_n sin(n\omega_0 t)) \tag{5.27}$$

where for C_n, $n \in \mathcal{Z}$ applies.

The representation of the Fourier series of a time function $x(t)$ is the amplitude spectrum for sine and cosine functions in the first case and the amplitude and phase spectrum for cosine functions in the second case. These spectra show, as in Eq. (5.27) the amplitudes A_n and B_n (or amplitude C_n and phase φ_n) for all $n \in \mathbb{N}$. Figure 5.17 shows the amplitude and phase spectrum for a given periodic signal $x(t)$.

As can also be seen in this figure, periodic signals have a line spectrum. The spectral lines have a fixed frequency spacing ω_0. It is easy to see that the line spacing is directly related to the period duration T_P: $T_P = 1/\omega_0$. Individual harmonics can

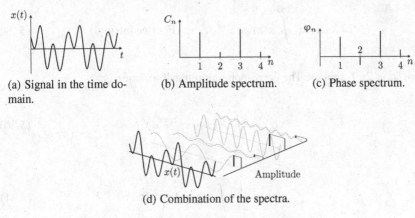

(a) Signal in the time do-main. (b) Amplitude spectrum. (c) Phase spectrum.

(d) Combination of the spectra.

Fig. 5.17 Signal with the corresponding amplitude and phase spectrum

be filtered out of the superimposed signal $x(t)$ by band filters, whereby the signal $x(t)$ is only approximated by omitting harmonics.

The shorter or steeper a pulse is in the time function $x(t)$, the more harmonics occur in the spectrum. This is particularly important for digital technology, since ideal square pulses have vertical edges and thus an infinitely wide spectrum. However, technical systems and real cables always show a low-pass effect, i.e. they dampen the harmonics more and more as the index increases. The more drastic this effect is, the more the square-wave signal, e.g. a clock signal, is deformed. This is taken up again below. Examples of Fourier series and their properties can be found in Appendix C.

The disadvantage of the description of signals by the spectra of the Fourier series is that only periodic signals can be described with the Fourier series. If one disregards the clock signal mentioned above, however, the practice-based signals of information technology are aperiodic and begin at a certain point t_0 in time. The Fourier series approach is no longer applicable here.

A mathematical trick is used to describe *aperiodic, finite signals*. The finite signal is regarded as a period of a periodic signal with an arbitrary period duration T_P. For consideration of the limit value $T_P \to \infty$, the following changes result in the Fourier series decomposition.

- The discrete line spectrum becomes a continuous spectrum (the lines grow together with increasing period duration T_P); this gives a continuous spectral function;
- the summation of the series decomposition is converted into an integration.

The relationship between the time signal $x(t)$ and the corresponding spectrum $X(f)$ is described by the Fourier transformation $\mathcal{F}\{x(t)\} = X(f)$.

$$X(f) = \int_{-\infty}^{\infty} x(t) \cdot e^{-j2\pi ft} dt \qquad \text{(Fourier-Integral)} \qquad (5.28)$$

or

$$X(\omega) = \int\limits_{-\infty}^{\infty} x(t) \cdot e^{-j\omega t} dt \qquad \text{(Fourier-Integral)} \tag{5.29}$$

Here, $\omega = 2\pi f$ applies. The corresponding retransformation is

$$x(t) = \int\limits_{-\infty}^{\infty} X(f) \cdot e^{j2\pi ft} df \tag{5.30}$$

or

$$x(t) = \frac{1}{2\pi} \int\limits_{-\infty}^{\infty} X(\omega) \cdot e^{j\omega t} dt \tag{5.31}$$

For practical reasons, instead of the cosine function, the equivalent complex notation based on the Euler relationship $e^{j\omega t}$ is used here usually.

The Fourier spectrum is generally complex. $X(f) = |X(f)| \cdot e^{j\varphi(f)} = Re(f) + j \cdot Im(f)$ and consists, as in the Fourier series, of an amplitude spectrum $|X(f)|$ and a phase spectrum $\varphi(f)$. For real-value signals the spectra show a certain symmetry. $|X(f)| = |X(-f)|$ and $\varphi(f) = -\varphi(-f)$. When converted, that means

$$Re(f) = Re\{X(f)\} = \int\limits_{-\infty}^{\infty} x(t) \cdot cos(2\pi ft) dt$$

and

$$Im(f) = Im\{X(f)\} = -\int\limits_{-\infty}^{\infty} x(t) \cdot sin(2\pi ft) dt$$

whereby the real part is an even function $[Re(f) = Re(-f)]$ and the imaginary part is an odd function $[Im(f) = -Im(-f)]$. Generally the following applies

- the transformed result of an even time function is purely real,
- the transformed result of an odd time function is purely imaginary.

Table 5.2 shows the relationship between the time function $x(t)$ and the amplitude spectrum $|X(f)|$ for some sample signals. Interesting in this context is the fact that the functions constant/Dirac impulse or rectangle/Si function have a dual relationship to each other in the time and frequency domain, i.e. they can be transformed into each other in both directions. In accordance with the series development, it is noticeable that signals which contain a steep rise in the time domain (such as the jump function or the square-wave signal) have a continuous frequency response with frequency components up to infinity.

The transformation of the Dirac delta function $\delta(t)$ (Eq. (5.32)) and the transformation of the square pulse $rect(t)$ (Eq. (5.33)) are regarded as examples.

Table 5.2 Selected functions with their corresponding Fourier spectra

Name	Time function	Amplitude spectrum
Constant, DC voltage	$x(t)$ t	$\|X(f)\|$ f
Dirac impulse	$x(t)$ t	$\|X(f)\|$ f
Jump function	$x(t)$ t	$\|X(f)\|$ f
Square function	$x(t)$ t	$\|X(f)\|$ f
Si function	$x(t)$ t	$\|X(f)\|$ f
Exponential pulse	$x(t)$ t	$\|X(f)\|$ f

$$X(f) = \int\limits_{-\infty}^{\infty} \delta(t) \cdot e^{-j\omega t}dt = e^0 = 1 \tag{5.32}$$

Table 5.3 Properties of the Fourier transformation

Prerequisite	$x(t)\circ\!\!-\!\!\bullet X(\omega)$
Linearity	$\mathcal{F}(a \cdot x + b \cdot g) = a \cdot \mathcal{F}(x) + b \cdot \mathcal{F}(g)$
Time shift	$x(a - t)\circ\!\!-\!\!\bullet e^{-ia\omega} \cdot X(\omega)$
Frequency shift	$e^{iat} \cdot x(t)\circ\!\!-\!\!\bullet X(\omega - a)$
Mirror symmetry	$\mathcal{F}(\mathcal{F}(x))(t) = x(-t)$
Convolution theorem	$\mathcal{F}(x * g) = (2\pi)^{\frac{n}{2}} \mathcal{F}(x) \cdot \mathcal{F}(g)$

$$X(f) = \int_{-\infty}^{\infty} rect(t) \cdot e^{-j\omega t} dt = \int_{-1/2}^{1/2} e^{-j2\pi f t} dt = \frac{-1}{j2\pi f}(e^{-j\pi f} - e^{j\pi f}) \quad (5.33)$$

$$= \frac{sin(\pi f)}{\pi f} = si(\pi f) = sinc(f)$$

with $si(x) = \frac{sin(x)}{x}$ and $sinc(x) = \frac{sin(\pi x)}{\pi x}$.

If we are dealing only with simple signals, or if the complex signal can be divided into a sum of elementary signals (e.g. by means of partial fraction decomposition or the residual theorem), the transformed signal can also simply be taken from a transformation table as can be found in Appendix C.2.

When working with the Fourier transformation, some mathematical properties have to be considered which can also serve as an aid. Here $\mathcal{F}(x)$ is the result of the Fourier transformation X of x (Table 5.3).

5.4.2 Discrete Fourier Transformation (DFT)

The Discrete Fourier Transformation is an important tool, especially in audio technology and when using digital signal processors, and it can be found in many codecs and transmission systems. Here the fact that digital signals are time-discrete is used, which is why the transformation can be represented by a sum instead of the integral, which simplifies its realization e.g. by a program. In contrast to the Fourier series, the Discrete Fourier Transformation can be used to transfer finite signals to the frequency domain. The Discrete Fourier Transform is defined as

$$X\left(\frac{k}{N \cdot T}\right) = T \cdot \sum_{n=0}^{N-1} x(nT) \cdot e^{-j \cdot 2 \cdot \pi \cdot \frac{k \cdot n}{N}}, k = 0, 1, \ldots, N - 1$$

N discrete function values → N discrete spectral values

$$x(n \cdot T), n = 0, 1, \ldots, N - 1 \rightarrow X\left(\frac{k}{N \cdot T}\right), k = 0, 1, \ldots, N - 1$$

Abbreviated form: $T = 1$, $X(k)$ instead of $X\left(\frac{k}{N \cdot T}\right)$:

$$X(k) = \sum_{n=0}^{N-1} x(n) \cdot e^{-j \cdot 2 \cdot \pi \cdot \frac{k \cdot n}{N}}, k = 0, 1, \ldots, N - 1 \qquad (5.34)$$

The corresponding *Inverse* Discrete Fourier transformation is

$$x(n) = \frac{1}{N} \sum_{k=0}^{N-1} X(k) \cdot e^{j \cdot 2 \cdot \pi \cdot \frac{k \cdot n}{N}}, n = 0, 1, \ldots, N - 1 \qquad (5.35)$$

Digital signal transmission and processing in particular benefit from the reduced computing complexity, which is reflected in lower latency times. In practice, this is further optimized by the use of *Fast Fourier Transformation* (*FFT*), which is a special case of the DFT.

The calculation rules known from the continuous Fourier transformation (see Sect. 5.4.1) can also be applied to the DFT (Table 5.4).

5.4.3 Laplace Transformation

A prerequisite for the Fourier transformation of a continuous signal mentioned above is the absolute integrability of the time function $x(t)$. However, for many important excitation functions of control engineering, such as the jump function or a time-limited linear increase, this does not apply. In this case, integration is achieved by multiplying the time function $x(t)$ by a damping function with $e^{-\alpha t}$ where $\alpha \in \mathcal{R}$.

Table 5.4 Properties of the DFT

Prerequisite	$\{x(n)\}_0^{N-1} \circ\!\!-\!\!\bullet \{X(k)\}_0^{N-1}$
Linearity	$\{a \cdot x(n)\}_0^{N-1} \circ\!\!-\!\!\bullet \{a \cdot X(k)\}_0^{N-1}$
Time shift	$\{x(n - n_0)\}_0^{N-1} \circ\!\!-\!\!\bullet \left\{ X(k) \cdot e^{-j \cdot 2 \cdot \pi \cdot \frac{k \cdot n_0}{N}} \right\}_0^{N-1}$
Frequency shift	$\left\{ x(n) \cdot e^{j \cdot 2 \cdot \pi \cdot \frac{k_0 \cdot n}{N}} \right\}_0^{N-1} \circ\!\!-\!\!\bullet \{X(k - k_0)\}_0^{N-1}$

$$X(f) = \int\limits_{0}^{\infty} x(t)e^{-\alpha t}e^{-j\omega t}\,dt = \int\limits_{0}^{\infty} x(t)e^{-(\alpha+j\omega)t}\,dt \qquad (5.36)$$

If now $s = \alpha + j\omega$ is used, the Laplace transform results to

$$x(t)\circ\!\!-\!\!\bullet X(s) = \int\limits_{0}^{\infty} x(t)e^{-st}\,dt \qquad (5.37)$$

or

$$\mathcal{L}\{x(t)\} = X(s) = \int\limits_{0}^{\infty} x(t)e^{-st}\,dt \qquad (5.38)$$

The corresponding inverse Laplace transform is

$$x(t) = \mathcal{L}^{-1}\{X(s)\} = \frac{1}{j2\pi} \int\limits_{c-j\infty}^{c+j\infty} X(s)e^{st}\,ds \qquad (5.39)$$

In the same way as the Fourier transformation, the Laplace transformation is also used to solve differential equations. It is particularly important in control engineering, since many of the modeling and identification methods (see Sect. 6.2.3.2) derive the system properties from the Laplace transform. It also clearly transfers a time function into a complex variable domain (also called Laplace domain or Laplace plane). Respectively, it maps a function with a real-value (time) variable to an image function with a complex-value (spectral) variable. It is often regarded as a generalization of the Fourier transform.

In practical applications, transformation tables also play an important role in the Laplace transformation, as can be seen in Table C.3. These tables can be used to transform simple components of an equation without additional computational effort. In practice, it is therefore often easier to divide an equation into a sum of matching terms by means of partial fraction decomposition, residual theorem, etc., and then to transform this with the table, than to calculate the transform of the complete equation directly. As known from the Fourier transformation, some calculation rules are also valid for the Laplace transformation, as shown in Table 5.5.

An important quantity in the Laplace transform, but also in the Fourier domain is the transfer function G. Since convolution in the time domain becomes multiplication in the frequency and complex variable domain, the calculation of the transfer function becomes

$$G(s) = \frac{Y(s)}{X(s)}$$

Table 5.5 Properties of the Laplace transformation

Linearity	$\mathcal{L}\{a_1 \cdot f_1(t) + a_2 \cdot f_2(t)\} = a_1 \mathcal{L}\{f_1(t)\} + a_2 \mathcal{L}\{f_2(t)\}$
Displacement law	$\mathcal{L}\{f(t-a)\} = e^{-as} \cdot \mathcal{L}\{f(t)\} = e^{-as} \cdot F(s)$ $\mathcal{L}\{f(t+a)\} =$ $e^{as} \cdot \left(F(s) - \int_0^a f(t)e^{-st}dt \right)$ for $t \geq a \geq 0$
Similarity law	$\mathcal{L}\{f(at)\} = \frac{1}{a} \cdot F(\frac{s}{a})$
Multiplication theorem	$\mathcal{L}\{t^n f(t)\} = (-1)^n \cdot F^{(n)}(s)$ for $n = 1, 2, 3, \ldots$
Differentiation theorem	$\mathcal{L}\left\{\left(\frac{d}{dt}\right)^n f(t)\right\}$ $= s^n \cdot F(s) - s^{n-1} f(0^+) - s^{n-2} f'(0^+) -$ $\cdots - f^{(n-1)}(0^+)$
Integration theorem	$\mathcal{L}\left\{\int_0^t f(q)dq\right\} = \frac{1}{s} \cdot F(s)$
Convolution theorem	$\mathcal{L}\{f_1(t) * f_2(t)\} = F_1(s) \cdot F_2(s) =$ $\mathcal{L}\left\{\int_0^t f_1(u) \cdot f_2(t-u)du\right\}$
Limit theorem	$\lim_{s \to 0} s \cdot F(s) = \lim_{t \to \infty} f(t) \qquad \lim_{s \to \infty} s \cdot F(s) = \lim_{t \to 0^+} f(t)$

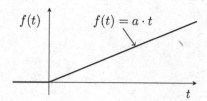

Fig. 5.18 Linear gradient

The transfer function thus represents the relationship between the input and output variables of a system.

As an example of the Laplace transformation, a linear gradient is considered, which is shown in Fig. 5.18.

The corresponding function is

$$f(t) = \begin{cases} a \cdot t & \text{for } t \geq 0 \\ 0 & \text{else} \end{cases}$$

By converting according to formula (5.37), this results in the Laplace transform

$$F(s) = a \int_0^\infty t \cdot e^{-st} dt = \frac{a}{s^2}$$

In practice, the transformation (if calculated manually) is carried out using tables (see Table C.3, no. 3).

$$a \cdot t \circ\!\!-\!\!\bullet \frac{a}{s^2}$$

References

1. Kossiakoff A, Sweet WN, Seymour SJ, Biemer SM (2011) Systems engineering principles and practice. Wiley, New Jersey
2. Chen WK (ed) (2004) The electrical engineering handbook. Elsevier, Burlington
3. Svoboda JA, Dorf RC (2013) Introduction to electric circuits. Wiley, Jefferson City

Chapter 6
Control Theory

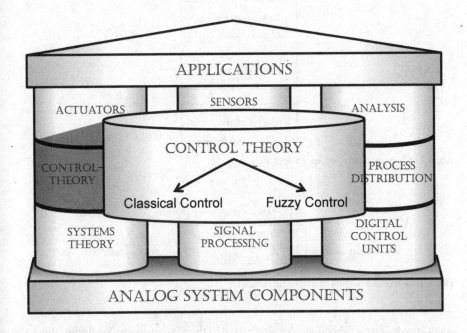

This chapter introduces the basic principles of control engineering according to the workflow shown in Fig. 6.1. First, the functional principle of a classical regulator and its design cycle are discussed. Important development and analysis tools are then presented, including block diagrams, step response measurements, Bode diagrams, locus curves and pole-zero diagrams. These are used, for example, for system identification. With the elementary members introduced next, it is possible to divide complex systems into easily analyzable components in order to determine their properties. Heuristic design methods are then introduced and the regulator design is explained

© The Author(s), under exclusive license to Springer Nature Switzerland AG 2021
K. Berns et al., *Technical Foundations of Embedded Systems*, Lecture Notes in Electrical
Engineering 732, https://doi.org/10.1007/978-3-030-65157-2_6

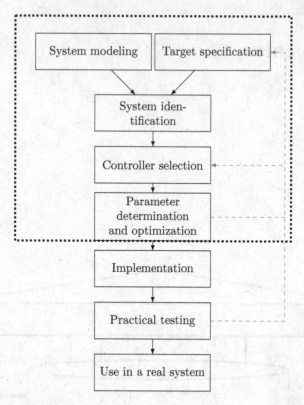

Fig. 6.1 Classical controller design approach

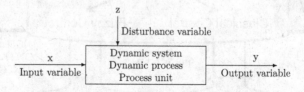

Fig. 6.2 Inputs and outputs of a dynamic system

according to Ziegler–Nichols. An alternative to the classic controller design, based on a more intuitive approach, is treated with fuzzy control.

For further information, we recommend for example [1–3].

6.1 Introduction to Control Engineering

As already shown in Chap. 1, reactive systems evaluate sensory information and act on the process via effectors/actuators. The sensor information provides one or

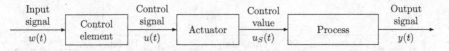

Fig. 6.3 Overview on the components of a open loop control chain

more regulating and disturbance variables (see Fig. 6.2). The system uses this data to attempt to execute the process as part of its specification.

Various control algorithms are used to achieve this. Their design and analysis are thus an integral part of the development of embedded systems.

6.1.1 Open-Loop Control

One intuitive way of influencing a process is the *open loop control* (also called *non-feedback control* or *feed-forward control*). It calculates the controlled variables without knowledge of the current actual state of the process. This requires a model that describes the relationship between controlled variables and output variables.

In this open chain, the input signal is given by a higher-level controller or by a user. This signal is converted into a control signal based on model knowledge of the process in the control element, as shown Fig. 6.3. For example, a gate valve with spindle drive is to be brought into a certain position. In this case, the input signal would be the position default. The control element calculates a period of time during which a voltage is applied that is amplified in the control element. This increased tension causes a rotation of the spindle and thus a movement of the slide (distance). In the event of a fault in the system, there is a deviation from the target value which is not detected in the control element. This means that no correction of the control is possible and the system could be destroyed in the worst case.

It should be noted that the term "control" is sometimes used differently in the literature. Particularly in industry, control is also used to describe the influence of a feedback system with discrete states (e.g. in the case of the PLC presented in Sect. 10.3.2). Here, more precise terms would be logical control, sequential control or discrete event control.

6.1.2 Closed Loop Control

The task of a controller is to change the input signal of the system in such a way that the output variable follows the setpoint setting as closely as possible. In contrast to an open loop control system, a closed loop controller monitors the output line via feedback in order to be able to react to faults. The controller receives the control difference e, resulting from the subtraction of the setpoint value w and feedback variable r as an input signal. An ideal controller adjusts the control difference to zero. A real controller tries to keep this difference as small as possible.

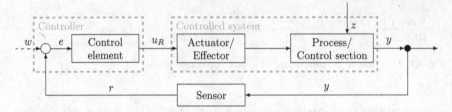

Fig. 6.4 Block diagram of a closed control loop

Closed loop control theory refers to the study of the automatic, targeted influencing of dynamic processes during the process flow. The methods of control engineering are generally valid, i.e. independent of the specific nature of the systems.

The following components are distinguished here (see Fig. 6.4):

- **Control section** The part of a system to be controlled.
- **Reference variable w** An externally supplied value which the output variable is to follow in a specified dependency.
- **Control difference e** Difference between the reference variable w and the feedback variable r.
- **Controller** Calculates an output variable from the control difference e, with the help of which the system will follow the reference variable w as quickly as possible.
- **Final control element/Actuator** Functional unit at the input of the system that intervenes in the mass flow or energy flow. The output variable is the controlled variable.
- **Controller output variable u_R** The input variable of the actuator.

In systems with analog quantities, the change due to interference is difficult to predict. For this reason, feedback controllers are predominantly used in these systems instead of simple controls. In contrast, in systems with binary or digital quantities, open loop controls are often used. The binary variables make the behavior more predictable.

6.1.3 Simple Controllers

The beginning of control engineering can no longer be determined exactly. Even in antiquity, mechanical controllers were used for various purposes, such as the level control of the Ctesibius of Alexandria. He had recognized that a water meter ran much more accurately if the water column of the storage vessel remained the same. He put a float valve in it. This consisted of a conical float pointing upwards, which moved in a cavity with the inflow at its upper end. If the water level rose, the float narrowed the inflow, if the water level fell, the inflow was opened again.

One of the first controllers of considerable economic relevance was the centrifugal pressure controller introduced by James Watt in 1788. As can be seen in the Fig. 6.5, this consisted of two weights attached to a shaft by movable arms. As the shaft

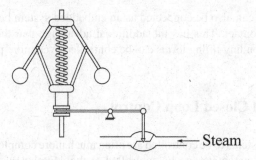

Fig. 6.5 Mechanical pressure controller

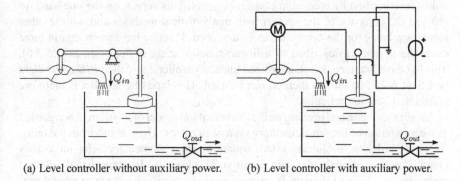

(a) Level controller without auxiliary power.   (b) Level controller with auxiliary power.

Fig. 6.6 Simple level controllers

rotated, the weights began to lift by centrifugal force, limited by a resetting spring and/or the weight force. This lifting caused a steam valve to close by means of a lever mechanism, so that less steam pressure arrived at the drive of the shaft, which thus rotated more slowly, which in turn caused the weights to lower, and so on until an equilibrium had been reached. Thus, the steam pressure and therefore the speed of the controller and a connected machine could be adjusted by pre-tensioning the spring (reference variable).

Another simple controller is the level controller without auxiliary power (Fig. 6.6a). This controller has a float in a container for a specific liquid. The float floats on the liquid and thus the (vertical) position of the float depends on the level. At a high level, the float limits or completely shuts off the inflow of liquid. If the liquid drains off, the filling level drops and the inflow of the liquid is opened again. The level controller with auxiliary power (Fig. 6.6b) works in a similar way. A sliding contact is attached to the float which regulates the voltage supply of a motor-controlled valve via a voltage divider. However, there is a problem here: if the water level is high enough or too high, the motor continues to close the valve with full force. Switching off at a sufficiently high water level would result in a more complex circuit. Instead, a processing unit can also be connected as an embedded system between the motor and the sliding contact. This has the additional advantage that the reaction of the motor to corresponding filling levels can be controlled even more precisely.

6.2 Classical Closed Loop Control

In practice, the systems to be controlled are often much more complex, either because of the number of parameters to be controlled or the complex system behavior. In order to find a meaningful control strategy, a mathematical model of these systems must therefore first be created in classical control. This serves, on the one hand, to analyze the behavior of the system with mathematical methods and, on the other hand, as a basis for the controller to be designed. Practice has shown that in most cases the system is described as a linear, time-invariant system (see Sect. 5.2.6). This has the advantage that standardized linear controllers, so-called *PID controllers* (see Sect. 6.2.5) or subsets thereof, can be used. This type of controller is therefore considered here in particular.

In order to obtain and investigate the mathematical model of a system, it is regarded as a compilation of known elementary system elements. These result from the terms of lowest order in the linear system equation. Their elementary behavior is thus clearly determined by their system equation. This has the advantage that all relevant quantities can be read directly from the equations or visualized via graphical representations. By means of the mathematically exact description concrete statements can be made regarding the behavior and the security and/or robustness of the controlled systems. Using appropriate methods, optimal solutions can also be found in line with the model.

6.2.1 Target Specification

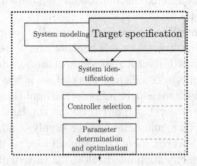

Before designing a controller for a system, you should be clear about the purpose of the controller. For this a rough knowledge of the system is necessary, but not necessarily a complete model. A number of formal quality criteria have been developed for this purpose, which will be examined in more detail below using the example of steering control for autonomous forklift trucks (see Fig. 6.4). For this, a so-called *Ackermann kinematic* model (a steerable front axle and a rigid rear axle, which corresponds to a normal car) is assumed for the chassis of the vehicle.

The obvious goal is for the vehicle to go where you want it to go, so that the actual steering angle matches what is required. This is referred to as

- **Steady-state error (Offset)** The controlled variable $y(t)$ should follow the reference variable $w(t)$ asymptotically. The remaining control difference is $e(\infty) = \lim_{t\to\infty} e(t) = \lim_{t\to\infty} (w(t) - y(t))$ (or $e(\infty) = \lim_{t\to\infty} (w(t) - r(t))$, if a measuring device is taken into account) and describes the control accuracy. If $e(\infty) = 0$, there is no residual steady-state error.

When driving on a road, the road surface is limited, but the vehicle continues to move and you do not want to get into the wrong lane. The steering angle required should therefore be achieved as quickly as possible. The

- **Dynamic control accuracy (speed)** Describes the time required by the controlled variable to react to changes in the reference variable.

The steering may have some play, or the road may be very uneven so that the desired steering angle cannot be achieved exactly. In this case, you may also be satisfied with a steering angle that deviates slightly from the one actually required and must therefore be corrected more often. So you have a certain

- **Tolerance range** A user-tolerated deviation from the nominal value in the steady-state condition.

It can also happen that the vehicle skids briefly (slippery roads, tight bends, etc.). In this case, the aim is to stabilize it as far as possible and to prevent it from rocking and possibly overturning. As with most controllers,

- **Stability** is required. This is affected negatively if the closed loop controller does not respond quickly enough to changes or cannot generate a control signal that counteracts the change. If the system becomes unstable, the feedback destroys the system. (Cf. Fig. 6.8).

In order to prevent the above case, the vehicle could be driven more slowly, for example, or other measures can be taken to protect it from rocking, so that there are still reserves in case of emergency. The aim here is to achieve a high degree of

- **Robustness** If a high quality of stationary and dynamic control and stability are achieved, the system is robust. Approximate models are used in the controller design for simplification.

Precise actual value acquisition, a fast actual/target value comparator, fast corrections and a small tolerance range form the basis for a good controller. A high dynamic and steady-state control quality is therefore required in order to work stably and robustly despite model uncertainties and unforeseeable disturbances.

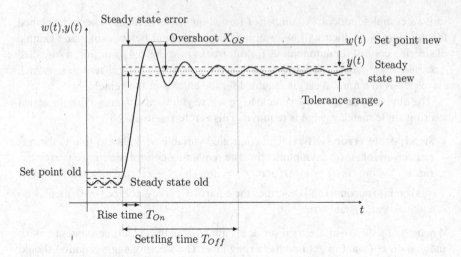

Fig. 6.7 Different measured parameters at a jump of the reference variable

The following parameters can be used to describe the dynamic control accuracy (see Fig. 6.7).

- **Rise time T_{On}** Minimum time span for moving from the tolerance range of a steady-state to another range for the first time.
- **Settling time T_{Off}** Time span for moving from the tolerance range of one steady-state to another conclusively.
- **Overshoot X_{OS}** Deflections of the controlled variable $y(t)$ over the target steady-state.
- **Control area I_R** Sum of all partial areas between $y(t)$ and the steady state $y(\infty)$ of a transition as a result of a disturbance or guidance jump. This is required for perfect control as a basis for integral criteria for controller design. (Optimization problem.)

The *stability* referred to above is one of the most important goals of control engineering and often a problem in controller design. This term will therefore be clarified further below:

- **Asymptotic stability** A linear system is asymptotically stable when it returns to its original state after a time-limited excitation.
- **Instability** A linear system is unstable if $|e(\infty)| = \infty$, so, if after an excitation, it does not remain limited.
- **Marginal stability** A linear system has marginal stability if it does not return to its original state after a time-limited excitation, but remains limited.
- **BIBO Stability** For each point in time t: Bounded Input $|x(t)| \leq N < \infty \Rightarrow$ Bounded Output $|y(t)| \leq M < \infty$. This definition of stability is called *BIBO stability* (Bounded *I*nput *B*ounded *O*utput). A linear, time-invariant transmission sys-

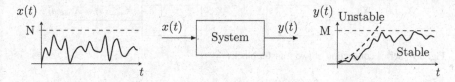

Fig. 6.8 BIBO stability in the time domain: output signal of a stable and an unstable system

tem is BIBO stable, if it is also **asymptotically stable**, cf. Fig. 6.8. (The reverse does not always apply!)

Unless explicitly mentioned otherwise, the terms *stability* and *stable* refer to BIBO stability below.

6.2.2   System Modeling and Identification

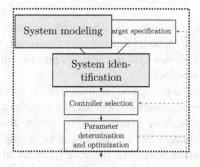

Once the target parameters of the system have been defined, the system at hand (the controlled system) without controller must be checked for these parameters before the actual controller design is carried out in order to determine the actual control requirements. If one considers, for example, the step response of a system given in Fig. 6.7, a residual steady-state error is evident. The system also overshoots, which is often undesirable. Furthermore, the control settling time is quite large compared to the control rise time, and so on. By comparing the knowledge gained in this way with the targets, it is now possible to determine which properties the controller must have. The way in which this knowledge is gained will be introduced in the following section.

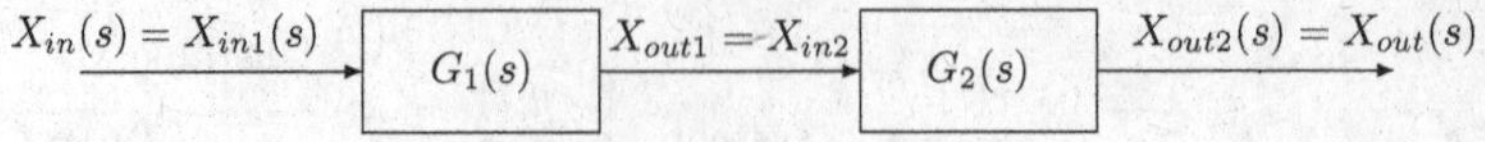

$$X_{in}(s) = X_{in1}(s) \xrightarrow{} \boxed{G_1(s)} \xrightarrow{X_{out1} = X_{in2}} \boxed{G_2(s)} \xrightarrow{X_{out2}(s) = X_{out}(s)}$$

Fig. 6.9 Series connection of two transfer blocks

The term *modeling* refers to the transformation of the properties of a real (or imaginary) system into an abstract representation. In practice, this usually means setting up a mathematical equation system that describes the relevant properties of the system, such as its physical behavior. By means of these equations, conclusions can be drawn formally about the behavior of the system in certain situations. This process is referred to as *identification*. In addition to the use of pure mathematical equations, graphical representation methods have also become established. These often offer an intuitive approach or allow quick conclusions to be drawn about system behavior. Some of the important methods of representation are presented in more detail below.

6.2.2.1   Block Diagrams

More complex dynamic systems can be illustrated very clearly by means of block diagrams. Several individual transfer functions are connected in series, in parallel and with feedback to more complex systems. For this purpose, these transfer functions are represented in discrete blocks.

This makes it possible to determine the total transfer function from the existing transfer functions of individual subcomponents of a complex system using standardized methods and to adapt these dynamically to system changes. The diagram is read as follows: What appears "in the block" can be a graphical or mathematical representation of the transfer function represented. Unless otherwise indicated, this is multiplied by the incoming signal. It should be noted that the display often takes place in the complex variable domain (Laplace domain). In many cases this simplifies the calculation. This can be recognized by the fact that variable sizes are represented by capital letters and by the occurrence of the "Laplace-s" (see Sect. 5.4.3).

For the analysis of these models, series connection, parallel connection and feedback connection can be calculated (decomposed).

The two series-connected (transfer) functions G_1 and G_2, as illustrated in Fig. 6.9, are calculated as

$$G_1 = \frac{X_{out1}}{X_{in1}}, G_2 = \frac{X_{out2}}{X_{in2}}$$

From this, the total transfer function can be calculated to

$$X_{out}(s) = X_{out2} = G_2 \cdot X_{in2} = G_2 \cdot X_{out1}$$
$$X_{out1} = G_1 \cdot X_{in1} = G_1 \cdot X_{in}(s)$$
$$X_{out}(s) = G_1 \cdot G_2 \cdot X_{in}(s) \tag{6.1}$$

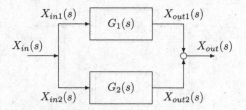

Fig. 6.10 Parallel connection of two transfer blocks

The transfer function of the subsystems connected in series G_1 and G_2 is therefore the product of the two individual functions: $G = G_1 \cdot G_2$

In general, functions G_n connected in series are multiplied.

$$G = \prod_n G_n \tag{6.2}$$

In Fig. 6.10, a parallel connection is shown. With parallel connection, the input signals are the same for both subsystems, the output signals are added together.

$$X_{in}(s) = X_{in1} = X_{in2}, X_{out}(s) = X_{out1} + X_{out2} \tag{6.3}$$

If the transfer functions G_1 and G_2 are inserted into these equations, the result for the parallel connection is that the transfer functions are added.

$$X_{out}(s) = G_1 \cdot X_{in1} + G_2 \cdot X_{in2} = (G_1 + G_2) \cdot X_{in}$$
$$\Rightarrow G = \frac{X_{out}}{X_{in}} = G_1 + G_2 \tag{6.4}$$

This equation can also be generalized for the parallel connection of several partial functions G_n.

$$G = \sum_n G_n \tag{6.5}$$

Feedback (Fig. 6.11) is regarded as the third basic type of circuit. This plays an important part in control engineering.

$$X_{out} = X_{out1} = G_1 \cdot X_{in1} \Rightarrow X_{in1} = \frac{1}{G_1} X_{out} \tag{6.6}$$

$$X_{out2} = G_2 \cdot X_{out} \tag{6.7}$$

$$X_{in1} = X_{in} \pm X_{out2} \overset{\text{with Figs. 6.6a, 6.7}}{\Longrightarrow} X_{in} = \left(\frac{1}{G_1} \pm G_2 \right) X_{out} \tag{6.8}$$

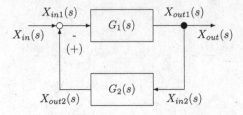

Fig. 6.11 Feedback of two transfer blocks

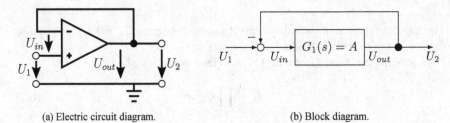

(a) Electric circuit diagram.   (b) Block diagram.

Fig. 6.12 Electric circuit and corresponding block diagram of an operational amplifier

For the transfer function of the total circuit, this results in

$$G = \frac{G_1}{1 \pm G_1 \cdot G_2} \tag{6.9}$$

By means of these basic rules even complex block diagrams can be compiled. This will now be described using a few examples.

An operational amplifier is a differential amplifier with extremely high gain, i.e. a minimum difference between U_1 and U_2 leads to the maximum output voltage (U_s). Due to the extremely high input resistance, the input currents can be considered to be close to zero. In order to determine the block diagram of this system, its transfer function must first be established. From this a relationship between the input and output variable can be determined, which is then converted into a block diagram.

In the circuit in Fig. 6.12a, without a component in the feedback the function is $G_2 = 1$. Thus the transfer function of the OpAmp itself equals $G_1 = A$. The result is the block diagram in Fig. 6.12b. For analysis of this circuit, A is still assumed to be very large. As known from the formula (6.9), the transfer function of the complete feedback system is

$$G = \frac{A}{1 + A \cdot 1} = \frac{1}{1 + \frac{1}{A}} \approx 1 \tag{6.10}$$

which means that $U_2 = U_1$. As the voltage at both ends of the circuit is the same, but (nearly) no current is drawn, the circuit achieves electrical separation of two partial circuits.

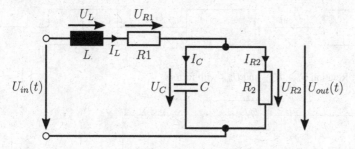

Fig. 6.13 Circuit diagram of a 2nd order low pass filter

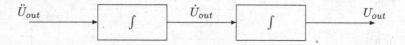

Fig. 6.14 Relation U_{out}, $\dot{U}_{out}$ and $\ddot{U}_{out}$

The low pass filter of Fig. 6.13 will be calculated as an example. Unlike the corresponding example in Sect. 3.5.1, it is a second order low pass due to the two energy stores, the capacitor and coil. Node and mesh equations give (in the time domain)

$$U_{out}(t) = U_{in}(t) - U_L(t) - U_{R_1}(t)$$
$$U_L(t) = L\dot{I}_L(t)$$
$$U_{R_1}(t) = I_L(t) \cdot R_1$$
$$I_L(t) = I_{R2} + I_C = \frac{U_{out}(t)}{R_2} + C \cdot \dot{U}_{out}(t)$$

It follows

$$U_{out}(t) = U_{in}(t) - L \cdot \frac{d}{dt}\left(\frac{U_{out}(t)}{R_2} + C \cdot \dot{U}_{out}(t)\right) - \left(\frac{U_{out}(t)}{R_2} + C \cdot \dot{U}_{out}(t)\right) \cdot R_1$$

It is useful to convert the differential equation to the highest derivative in order to create the block diagram. As L, R_2 and C are constant, the result is

$$\ddot{U}_{out}(t) = \frac{1}{LC}\left[U_{in}(t) - \dot{U}_{out}(t)\left(\frac{R_1 R_2 C + L}{R_2}\right) - U_{out}(t) \cdot \frac{R_1 + R_2}{R_2}\right]$$

It is now advisable to draw in the derivations of U_{out} (see Fig. 6.14).

Then it becomes apparent that the whole circuit is multiplied by $\frac{1}{LC}$ to be equal to $\ddot{U}_{out}$ (Fig. 6.15).

The remainder of the equation is an addition of U_{in}, $\dot{U}_{out}$ and U_{out} with the respective pre-factor (Fig. 6.16).

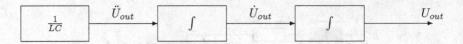

Fig. 6.15 Relation U_{out}, $\dot{U}_{out}$ and $\ddot{U}_{out}$ with factors

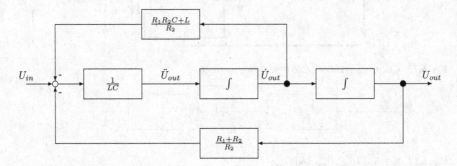

Fig. 6.16 Block diagram of the low pass in the time domain

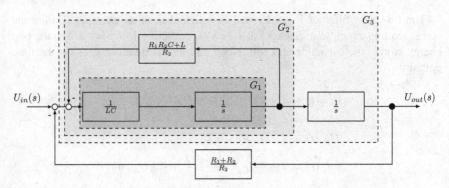

Fig. 6.17 Block diagram of the low pass in the complex variable domain

The total transfer function (see Fig. 6.17) is calculated in the complex variable domain (or Laplace domain, see Sect. 5.4.3). It follows from this that integration becomes multiplication by $\frac{1}{s}$ and differentiation becomes multiplication by s.

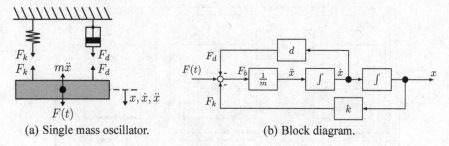

(a) Single mass oscillator.   (b) Block diagram.

Fig. 6.18 Sketch of the single mass oscillator and the corresponding block diagram

$$G_1 = \frac{1}{LC \cdot s}$$

$$G_2 = \frac{G_1}{1 + \frac{R_1 R_2 C + L}{R_2} \cdot G_1} = \frac{R_2}{LCR_2 s + R_1 R_2 C + L}$$

$$G_3 = G_2 \cdot \frac{1}{s} = \frac{R_2}{LCR_2 s + R_1 R_2 C + L} \cdot \frac{1}{s} = \frac{R_2}{LCR_2 s^2 + (R_1 R_2 C + L)s}$$

$$G_{tot} = \frac{U_{out}(s)}{U_{in}(s)} = \frac{1}{\frac{1}{G_3} + \frac{R_1 + R_2}{R_2}} = \frac{R_2}{LCR_2 s^2 + (R_1 R_2 C + L)s + R_1 + R_2}$$

(cf. Eq. (6.9))

$\Rightarrow$ Laplace transform of the differential equation of the RLC network :

$$(LCR_2 s^2 + (R_1 R_2 C + L)s + R_1 + R_2)U_{out}(s) = R_2 U_{in}(s)$$

In order to check the procedure and to be able to make statements about the temporal behavior of the circuit again, the received transfer function must be transformed back into the time domain. As a time function it is (see transformation rules in Table 5.5)

$$LCR_2 \cdot \ddot{U}_{out}(t) + (R_1 R_2 C + L) \cdot \dot{U}_{out}(t) + (R_1 + R_2) \cdot U_{out}(t) = R_2 \cdot U_{in}(t)$$

The methodology of block diagrams can, of course, be applied not only to electrical systems, but also to any other system. A single mass harmonic oscillator (Fig. 6.18) is considered below. The single mass oscillator with the body of the mass m is suspended from a spring F and a damper D. If m is deflected by the force $F(t)$ by the distance x, the spring force F_k and the damping F_d counteract $F(t)$. The deflection x of the mass is to be regulated. The reference variable is the force $F(t)$.

$$F_b = F(t) - F_k - F_d = m \cdot \ddot{x} \qquad (6.11)$$

Spring and damper forces then result to

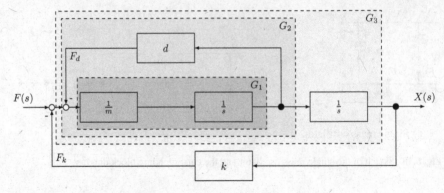

Fig. 6.19 Laplace transformed block diagram of the single mass oscillator

$$F_k = k \cdot x \tag{6.12}$$

$$F_d = d \cdot \dot{x} \tag{6.13}$$

After insertion and conversion, the highest derivative of the controlled variable is obtained.

$$\ddot{x} = \frac{1}{m}\left(F(t) - k \cdot x - d \cdot \dot{x}\right) \tag{6.14}$$

The resulting block diagram is shown in Fig. 6.18b on the right, the Laplace transformation in Fig. 6.19.

The overall transfer function is calculated by adding together the partial transfer functions.

$$G_1 = \frac{1}{m \cdot s} \tag{6.15}$$

$$G_2 = \frac{G_1}{1 + d \cdot G_1} = \frac{1}{m \cdot s + d} \tag{6.16}$$

$$G_3 = G_2 \cdot \frac{1}{s} = \frac{1}{m \cdot s + d} \cdot \frac{1}{s} = \frac{1}{m \cdot s^2 + d \cdot s} \tag{6.17}$$

$$G_{tot} = \frac{G_3}{1 + G_3 k} = \frac{1}{\frac{1}{G_3} + k} = \frac{1}{ms^2 + ds + k} = \frac{X(s)}{F(s)} = \frac{X_{out}(s)}{X_{in}(s)} \tag{6.18}$$

By solving the equation with $X_{in}(s)$, the result is

$$(ms^2 + ds + k) \cdot X_{out}(s) = ms^2 \cdot X_{out}(s) + ds \cdot X_{out}(s) + k \cdot X_{out}(s) = X_{in}(s) \tag{6.19}$$

Retransformation into the time domain finally gives the overall transfer function

$$m \cdot \ddot{x}(t) + d \cdot \dot{x}(t) + k \cdot x(t) = F(t) \tag{6.20}$$

6.2.3   Identification

The block diagram presented above is a good tool for clarifying interrelationships in a system and simplifying mathematical calculations. In this respect, it can be used for modeling purposes. However, in order to design a controller for a system, statements about the behavior and properties of the system must be obtained from a model. Only then is it possible to see exactly what the controller has to do. For example, in order to determine the speed of the controller, knowledge of system dynamics and possible dead times is required. The procedure used to obtain this information is called *identification*. Some of the methods used for this purpose are presented below.

6.2.3.1   Step Response

A basic means of system description and analysis is the step response. Here a system is excited with a step of the reference variable from 0 to 1 (mostly minimum deflection to maximum deflection) and then the way on which the output variable reacts to it is examined. Based on this analysis, which takes place in the time domain, it is immediately apparent whether there is a residual steady-state error (is the target value reached?), how fast it is (control rise and settling time), whether the system exhibits dead time behavior (does the system react directly to the step?) and much more. The step response results mathematically from the convolution of the unit step σ_t with the transfer function of the system in the time domain or from the multiplication of the respective Laplace transform in the complex variable domain.

If, for example, the step response in Fig. 6.20 is considered, several statements can be made without further knowledge of the underlying system. On the one hand, you can see that the system is vibratory because it has pronounced overshoots. There is a steady-state error, as the step response never reaches the required final value of 1, but settles at 0.8. The system reacts directly (it changes at time $t = 0$), so it seems to have no significant dead time. The control rise time is about 2.5 s (assuming a final value of 0.8), the settle time is about 25 s, depending on the tolerance range. Furthermore, the system seems to strive for a fixed end value $<\infty$, so it is stable. Thus, important knowledge about the system has already been gained.

6.2.3.2   Frequency and Complex Variable Domain

As shown in Chap. 5, the description of a system in the frequency or complex variable domain not only allows simpler calculations of the transfer function, but also an

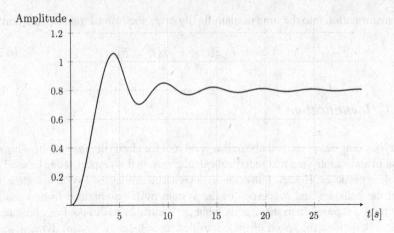

Fig. 6.20 Example of a step response

additional view of the system behavior. The Fourier transform of a signal $x(t)$, given by (see Sect. 5.4.1)

$$X(j\omega) \equiv \text{FT}\{x(t)\} = \int_{-\infty}^{\infty} x(t) \cdot e^{-j\cdot\omega t} dt \qquad (6.21)$$

establishes the amplitude and phase spectrum via polar form

$$X(j\omega) = |X(j\omega)| \cdot e^{j\cdot\varphi(\omega)} \qquad (6.22)$$

to

$$|X(j\omega)| = \sqrt{Re(\omega)^2 + Im(\omega)^2} \qquad \text{Amplitude spectrum}$$

$$\varphi(\omega) = \arctan\left(\frac{Im(\omega)}{Re(\omega)}\right) \qquad \text{Phase spectrum.}$$

The amplitude and phase curves of a transfer function in the frequency domain obtained in this way can be used to make direct statements about the behavior of the system. In practice, however, the Fourier transform representation has proved impractical for the calculation of poles and zeros in phase response, since absolute integrability of the test signals used is not always given. Therefore, the Laplace transform is usually chosen in control engineering, which is less problematic here due to the complex frequency s.

By means of the Laplace transformation familiar from Sect. 5.4.3, a frequency-dependent representation can also be achieved, which is referred to here as the complex variable domain.

6.2.3.3 Bode Plot

The amplitude and phase spectra of a system transfer function provide important information about the behavior of the system under investigation, in particular about its stability, robustness and behavior in response to excitation at different frequencies. A *Bode plot* can be used to visualize this.

Amplitude and phase response are plotted over the logarithmically divided frequency or circular frequency. The amplitude spectrum is converted into decibels as an *amplitude response*.

$$|X(j\omega)| \ \hat{=} \ 20 \cdot \log(|X(j\omega)|) \ [\text{dB}] \tag{6.23}$$

Here, $j\omega$ corresponds to the complex frequency s (the damping term σ is omitted in this representation). Also note, that in the following the usage of *root power quantities* (such as voltage or current) as controlled variables is assumed, so that the calculation of decibel includes the factor 20 instead of 10. If a power quantity was to be controlled, the "normal" equation $|X(j\omega)| \ \hat{=} \ 10 \cdot \log(|X(j\omega)|) \ [\text{dB}]$ would apply. The logarithmically plotted frequency and the converted amplitude response make it possible to display wide ranges of transmission behavior clearly. The value of amplitude and phase response can be presented as a system response to an excitation with a sinusoidal oscillation of the corresponding frequency. The amplitude response shows the (circular) frequency-dependent amplification of the input signal and the phase response the phase shift with respect to the input signal.

The Bode plot of a high-pass filter (Fig. 6.21) is used as an example. The amplitude response clearly shows that signals with frequencies of 100 $\frac{1}{s}$ or more are let through with a gain of [0] dB (corresponds to factor 1), while the frequencies below are suppressed. This corresponds to a 1st order high pass with the marginal frequency $f_G = 15.9$ Hz. The angular frequency is not given in *Hertz*, but in $\frac{1}{s}$, for better distinction. The phase response shows that the phase shift decreases with increasing frequency compared to the original signal.

Furthermore, the Bode plot can be used for stability estimates of feedback systems. The rule of thumb is: If the phase shift of an open-loop system is $<180°$ when crossing the 0 dB line of the amplitude response, the closed-loop system is stable.

In practice, this usually means that the phase at 0 dB must be greater than $-180°$. This is based on the *simplified Nyquist stability criterion*, which we will pick up below. If the amplitude response always remains below 0 dB or the phase response always above $-180°$, the system is also stable. Since there is no explicit transgression of 0 dB with the high pass filter but this value is reached, the phase should always be $>-180°$ from 100 $\frac{1}{s}$, which is given. So this system would remain stable, if fed back. This is more visible in the following abstract system. The transfer function is

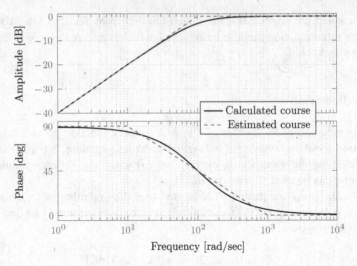

Fig. 6.21 Example: Bode plot of a high pass filter

given in the frequency domain.

$$X(j\omega) = \frac{1}{j\omega \cdot (1 + j\omega)} = \frac{1}{j\omega - \omega^2}$$

$$= -\frac{1}{1 + \omega^2} - j\frac{1}{\omega \cdot (1 + \omega^2)}$$

$$|X(j\omega)| = 20 \cdot \log\left(|X(j\omega)|\right) \, [\text{dB}]$$

$$\varphi(j\omega) = \arctan\left(\frac{1}{\omega}\right)$$

The resulting Bode plot can be seen in Fig. 6.22. At the frequency ($\omega = 10^0\frac{1}{s} = 1\frac{1}{s}$), the phase is at $-135°$. Thus the system would remain stable, if fed back.

Additional information about the *phase reserve* can be given here. It amounts to $|-180° - (-135°)| = 45°$. The phase reserve is a criterion for the robustness of a system to external influences. As a rule of thumb, it can be said that a system with a phase reserve of $>=30°$ is sufficiently robust. Alternatively, the *amplitude reserve* can also be used for this purpose. This is calculated from the value of the amplitude response with a phase shift of $-180°$.

6.2.3.4 Nyquist Plot

If the frequency response $G(j\omega)$ is represented as a curve in the complex plane as a function of the angular frequency ω, this is referred to as the *locus* or the *Nyquist plot* of a system. The Nyquist plot is an equivalent representation of the

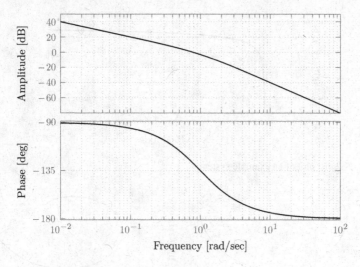

Fig. 6.22 Example of the Bode plot of a low pass filter

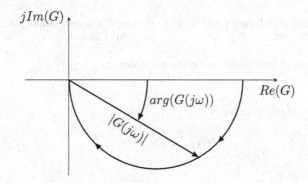

Fig. 6.23 Nyquist plot of a first order system

Bode plot. While the Bode plot emphasizes the frequency behavior of a system, this information is more compressed in the Nyquist plot. It has some advantages when considering system stability, especially for more complex systems. The amplitude response $AR(j\omega) = |G(j\omega)|$ and the phase response $\varphi(j\omega) = \arg(G(j\omega))$ are also considered here. In contrast to the Bode plot, however, these are not separated but drawn as a transfer function in the complex plane (see Fig. 6.23).

Important points are the amplification at $\omega = 0$ (starting point of the curve), at $\omega = \infty$ and at the *marginal frequency*. This is the frequency at which the amplitude response has decreased to $\frac{1}{\sqrt{2}}$ of its start value (equivalent to –3 dB).

Consider the system with the transfer function

$$G(j\omega) = \frac{1}{j\omega \cdot (1 + j\omega)}$$

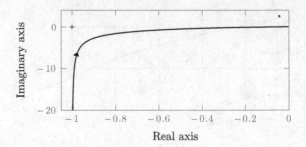

Fig. 6.24 Nyquist plot of an example system

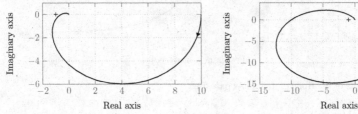

(a) Stable system, critical point (-1;0) remains   (b) Unstable system, critical point (-1;0) is on
on the left of the curve.   the right of the curve.

Fig. 6.25 Nyquist plots of a stable and an unstable system

the starting point of the curve is at

$$|G(j\omega = 0)| \to +\infty$$

and the endpoint at

$$|G(j\omega \to +\infty)| \to 0$$

Thus the curve runs along

$$\arg(G(j\omega)) : -90° \to -180°$$

(see Fig. 6.24).

The Nyquist plot with the *critical point* offers a simple way to examine stability. This point at −1 on the real axis represents a phase shift of (−)180° at a gain of 1 (0 dB). According to the stability criterion in the Bode plot, it can therefore be said in simplified terms that the (feedback) system is stable if the curve (of the open-loop system) does not "revolve" around point −1, i.e. if it passes to the right of it (see Fig. 6.25). This rule is called the *simplified Nyquist stability criterion*. (Not to be confused with the Nyquist–Shannon sampling theorem, which will be covered in Sect. 7.3.1.)

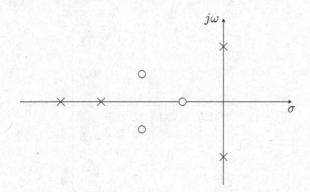

Fig. 6.26 Characterization of a linear system by the position of its poles (x) and zeros (o)

6.2.3.5 Pole-Zero Plot

Another relevant form of representation is the pole-zero plot. Here the pole and zero points of a transfer function in the complex variable domain are plotted in the complex number plane, as can be seen in Fig. 6.26. This method enables the stability of a control loop to be determined quickly. If this diagram is extended to a so-called *root locus*, the occurrence of oscillations can be investigated easily and it can also be used to determine the control parameters. However, this would go beyond the scope of this chapter, which is why reference should be made here to specialist literature (e.g. [1]).

As the name suggests, the pole-zero plot is made up of the pole and zero points of a transfer function:

- **Poles** of signals are present when their amplitude density assumes infinitely high values in the frequency domain. They are determined by the zero points of the denominator term of their transfer functions. In the diagram they are represented by crosses.
- **Zeros** of signals or their functions are defined in the frequency domain by the zeros of the corresponding numerator term. They are marked as circles on the diagram.

To span the complex number plane, the components of the Laplace variable $s = \sigma + j\omega$ are used. As an example, the following function is examined (see Fig. 6.27).

$$G(s) = \frac{s+2}{s^2+s}$$

with $p_1 = 0$, $p_2 = -1$ and $n_1 = -2$.

The pole-zero plot provides an easy and fast way to make accurate statements about the stability of a system:

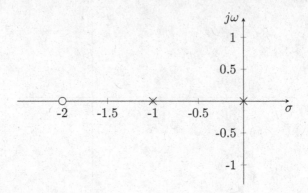

Fig. 6.27 Pole-zero plot example

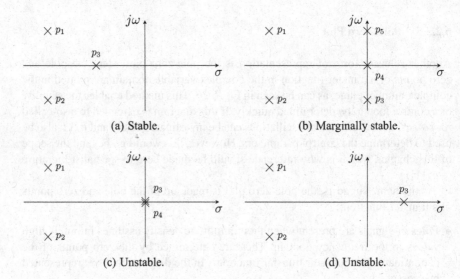

Fig. 6.28 Stability in the complex variable domain

- **Asymptotic stability** A linear system is asymptotically stable if the poles p_i are all in the left half-plane: Re $(p_i) < 0$ for $i = 1, 2, \ldots, n$ (Fig. 6.28a).
- **Instability** A linear system is unstable if at least one pole lies on the right's half-plane (Fig. 6.28d) or if at least one multiple pole is present on the imaginary axis. (Multiple pole means that both real part and imaginary part of several poles are equal, Fig. 6.28c.)
- **Marginal stability** A linear system is marginally stable if there is no pole on the right's half-plane and no multiple poles, but at least one single pole on the imaginary axis (Fig. 6.28b).
- A linear, time-invariant transmission system is **BIBO stable**, if it is also **asymptotically stable**.

Fig. 6.29 Overview on the
poles (crosses) in the
complex number plane.
Poles determine the course
of the system response

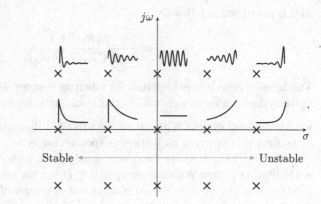

An overview of the oscillation behavior of a system depending on the position of its poles is provided in Fig. 6.29.

The tools and methods of modeling and identification presented will now be summarized using an example. The low-pass filter, whose block diagram has already been presented in Sect. 6.2.2.1, is used again for this purpose. Thus its total transfer function is also known. In the Laplace domain, it is

$$(LCR_2s^2 + (R_1R_2C + L)s + R_1 + R_2)U_{out}(s) = R_2U_{out}(s) \qquad (6.24)$$

It is a second order system (s^2 corresponds to second derivative in the time domain). As will be explained later, such systems are in principle capable of oscillating (see Sect. 6.2.4.4). It is therefore appropriate to bring the equation into the characteristic form for second-order systems

$$G = \frac{K}{1 + 2DTs + T^2s^2}$$

using K: gain, T: time constant and D: damping. So for the low pass, the equation is

$$G_{tot} = \frac{U_{out}}{U_{in}} = \frac{1}{LC} \cdot \frac{1}{s^2 + \frac{R_1R_2C+L}{LCR_2}s + \frac{R_1+R_2}{LCR_2}} = \frac{\frac{R_2}{R_1+R_2}}{\frac{LCR_2}{R_1+R_2}s^2 + \frac{R_1R_2C+L}{R_1+R_2}s + 1}$$

From this, some variables of the system can be directly read by means of parameter comparison. Thus the angular frequency of the undamped system can be determined from the time constant.

$$\frac{1}{T} = \omega_0 = \sqrt{\frac{R_1 + R_2}{LCR_2}}$$

this corresponds to the frequency with which the system would oscillate without further excitation if it were not damped. Thus the value of the damping is of interest.

This is calculated as follows.

$$D = \frac{1}{2} \frac{R_1 R_2 C + L}{\sqrt{L C R_2 (R_1 + R_2)}}$$

The damping term is very important for vibrating systems, as the vibration behavior of the system can be seen directly by this term. There are three different cases:

- **Overdamped system** With a damping of $D \geq 1$, the system slowly approaches the final value without oscillation. A special case is the *critical damping* with a damping of $D = 1$, where the system is only just not oscillating.
- **Oscillating system** With a damping of $0 \leq D < 1$ the system is capable of oscillation, whereby the case $D = 0$ is described as *marginally stable*, since here the system neither increases or decreases in oscillation.
- **Instability** When damped with $D < 0$, the system oscillates with increasing amplitude and therefore, it is unstable.

To illustrate this, imagine a pendulum that is completely deflected once and then released. In an empty space with no air or friction on the suspension, it would always continue to oscillate at the same frequency f. The angular frequency corresponds to $\omega = 2\pi f$. In the real world, the pendulum is damped by air friction and friction on the suspension, i.e. its amplitude is reduced over time until it no longer oscillates. This reduction corresponds to the damping.

- With a damping of $D >= 1$, the pendulum would be so heavily damped that it would not overshoot a single time, but would only move from the maximum deflection directly to the rest point.
- With a damping of $0 < D < 1$, the pendulum would oscillate, but the oscillation would become smaller and smaller and finally stop.
- With a damping of $D = 0$, the pendulum would simply continue swinging with the same amplitude forever.
- With a damping of $D < 0$, the pendulum would swing out increasingly, in the worst case to the point of destruction. This could happen, for example, by someone pushing it further and further.

For the low-pass filter, therefore, three cases are shown in Table 6.1 as examples.

Here the methods presented above are used to identify the system more precisely. The step response is considered first (Fig. 6.30).

Several statements can be made from this. The first system (G_1) has no overshoot, as is to be expected with damping of $D > 1$. The other two systems have overshoots with damping $D < 1$. The step responses of the two systems G_1 and G_2 have the same end value, but in both cases, a steady-state error remains. It can be concluded that R_1 and R_2 influence the final value. The next step is to examine the Nyquist plot.

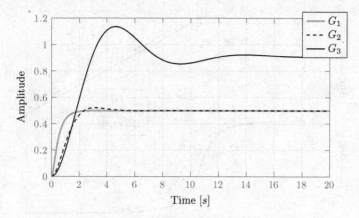

Fig. 6.30 Step response of the low pass in the time domain

Table 6.1 Exemplary values of the low pass filter for different damping values

Parameter	G_1	G_2	G_3
R_1	1	1	1
R_2	1	1	10
C	0.1	1	1
L	1	1	2
Damping	>1	0.707	0.405

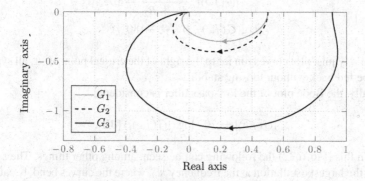

Fig. 6.31 Nyquist plot of the low pass filter

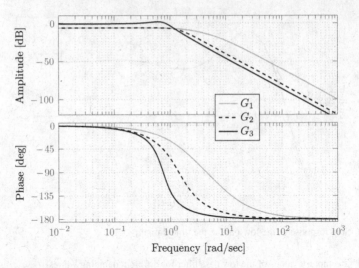

Fig. 6.32 Bode plot of the low pass filter

To determine the Nyquist plot (Fig. 6.31), the starting point, the end point and the angle progression must be known.

$$\omega : 0 \rightarrow +\infty$$
$$|G(j\omega)| : \frac{R_2}{R_1 + R_2} \rightarrow 0$$
$$\arg(G(j\omega)) : 0° \rightarrow -180°$$

Since all Nyquist plot curves run far to the right of the critical point at −1, all systems could be fed back without loosing stability.

Finally, the Bode plot of the low-pass filter is considered.

$$|G(j\omega)| = 20 \cdot \log\left(|G(j\omega)|\right) [\text{dB}].$$

From this (Fig. 6.32) the following can be seen, among other things. The system G_3 has the largest oscillation at the frequency w_0, where the curves bend, because its damping of <1 is the smallest. The system G_1 has no oscillation due to a damping of >1. The system G_3 still has a phase reserve of well over 30° for the (second) 0 dB pass, meaning that it is also robust.

The pole-zero diagram of the low-pass filter also provides us with information about the system. The system G_1 has two poles on the real axis in the left half-plane (see Fig. 6.33). The system itself is therefore stable and does not oscillate. The systems G_2 and G_3 have imaginary poles, so these systems are able to oscillate. Since all poles are on the left half-plane (real part < 0), these systems are also stable.

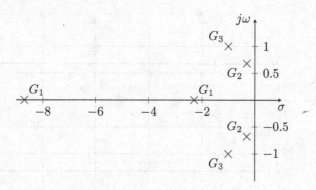

Fig. 6.33 Pole-zero plot of the low pass filter

6.2.4 Elementary System Elements

When creating block diagrams, one of the basic methods of classical control engineering has already been implicitly applied: the division of a complex overall system into individual elements with a known effect.

This procedure plays an important role in modeling and identification, especially as it enables quick estimation of system properties without complicated calculations or additional visualizations. This is based on the fact that the properties of the standard individual elements are already known. Since these effects overlap in the overall system in accordance with the block diagram, the individual elements can often be used to draw conclusions about the overall system.

The actual controllers are also regarded as a combination of such individual elements (see Sect. 6.2.6). Of course, the individual elements can represent any functions, but in classic controller design only a few standard *elementary elements* and composite standard elements are considered. Some of the important elementary elements and some composite standard elements are presented below and their properties for controller design are highlighted.

6.2.4.1 Proportional Element (P Element)

P elements are delay elements with proportional behavior. This means that after a change of the input x_{in} the output x_{out} strives for a new steady-state value. Even if pure P-links do not occur in reality, some real components can be described sufficiently precisely as such. Levers, voltage dividers (see Fig. 6.35 and cf. Sect. 3.2.1), amplifiers (see Sect. 4.9) and gears (see Sect. 9.4) are such examples of P elements (Fig. 6.34).

The transfer function of a P element is calculated from $x_{out}(t) = K_P \cdot x_{in}(t) \circ\!\!-\!\!\bullet K_P \cdot X_{in}(s)$ to

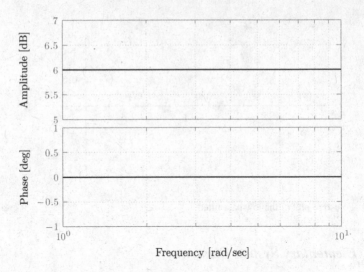

Fig. 6.34 Bode plot of a P element ($K = 2$)

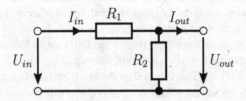

Fig. 6.35 Circuit implementation of a P element

$$G(s) = \frac{X_{out}(s)}{X_{in}(s)} = K_P \qquad (6.25)$$

Used as a controller, this results in $u(t) = K_P \cdot e(t)$. For the P element, the transfer function is $G(j\omega) = K$. Therefore, the amplitude and phase response in the Bode plot (cf. Fig. 6.34) applies.

$$|G(j\omega)| = K$$

$$\varphi(\omega) = 0$$

Properties of the P element:

- P elements are not capable of oscillating on their own.
- P elements do not reject a step disturbance, so when used as controllers they leave a *residual steady-state error*.
- For $K < \infty$ P elements are stable, but in combination with other elements they can lead to instability.
- As controllers, P elements react very quickly to a change in the controlled variable.

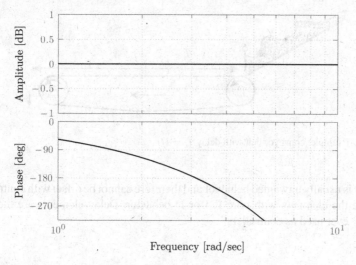

Fig. 6.36 Bode plot of a delay element ($K = 1$, $T = 1$)

6.2.4.2 Delay Element

The delay element (or dead time element) is a time delay in the signal. This can be seen from the fact that a system does not react directly to a change in the input, but takes a certain amount of time to react. Dead time behavior occurs, for example, in most transport tasks (conveyor belts, as shown in Fig. 6.37, in gear backlash between gears or in the form of the time of sensor data acquisition and processing. Mathematically, dead time behavior means that the input signal x_{in} with the delay T_t appears at the output of the system x_{out} with a delay. Thus the transfer function is calculated from $x_{out}(t) = x_{in}(t - T_t) \circ\!\!-\!\!\bullet X_{out}(s) = e^{-T_t \cdot s} \cdot X_{in}(s)$ to (Fig. 6.36)

$$G(s) = e^{-T_t \cdot s} \tag{6.26}$$

The amplitude response of the dead time element is $|G(jw)| = 1 \stackrel{\wedge}{=} 0\,\mathrm{dB}$ (see Fig. 6.36). The phase response of the dead time element is $\varphi(\omega) = -\omega \cdot T_t$.
  Properties of the dead time element:

- Delay elements are not capable of oscillating on their own, but can achieve this in combination with other elements.
- Delay elements leave a residual steady-state error, when used as a closed-loop controller.
- The stability of delay elements in a closed-loop system depends on the previous amplification. Since the phase shift increases with frequency (see phase response of the Bode plot), however, they tend to show unstable behavior and often bring this about in combination with other elements. In open loop systems, they are stable.

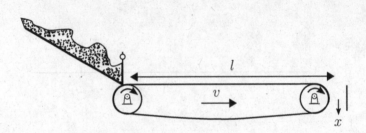

Fig. 6.37 Example: conveyor belt with delay $T_t = l/v$

- Delay is usually unwanted behavior and therefore cannot be offset with controllers, unlike the elements with pure P, I or D behavior. Delay elements are therefore hardly ever used as controllers.

6.2.4.3 P-T$_1$ Element

P behavior with 1st order delay is referred to as $P\text{-}T_1$ (Fig. 6.38).

An ideal P element without delay is not possible in practice, since an infinitely strong amplification would be necessary to follow an input step. The proportional feedback of an I element produces the P-T$_1$ element. This combination is very often found in practice, e.g. when charging a capacitor via a resistor (RC element, see Fig. 6.39, cf. Sect. 3.5). The output of a P-T$_1$ element generally approaches the steady-state value without overshoot with an e-function.

$$x_{out} = K \left(1 - e^{-\frac{t}{T}} \right) \sigma(t)$$

Here $\sigma(t)$ stands for the step function (see Sect. 5.2.4). After 3–5 times the time constant T, the P-T$_1$ element has reached 95% or respectively 99% of the steady-state value. The parameters K and T can, for example, be determined from the step response in the time domain.

The corresponding equation in the time domain is $T \cdot \dot{y}(t) + y(t) = K \cdot u(t)$. This results in the transfer function

$$G(s) = \frac{K}{1 + T \cdot s} \tag{6.27}$$

For the amplitude and phase response in the Bode plot (see Fig. 6.38), the result is therefore

$$|G(j\omega)| = \frac{K}{\sqrt{1 + \omega^2 T^2}} \tag{6.28}$$

$$\varphi(\omega) = -\arctan(\omega T) \tag{6.29}$$

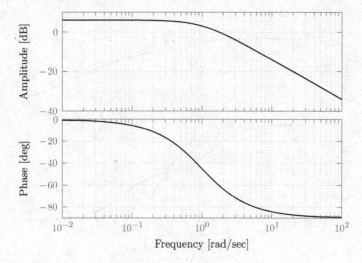

Fig. 6.38 Bode plot of a P-T$_1$ element ($K = 2$, $T = 1$)

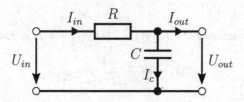

Fig. 6.39 Circuit implementation of a P-T$_1$ element

Properties of the P-T$_1$ element:

- The P-T$_1$ element is not capable of oscillating in itself.
- The P-T$_1$ element leaves a residual steady-state error, since the final value is theoretically reached only in infinity, but approaches it asymptotically.
- The P-T$_1$ element is stable in an open loop system for $T \geq 0$, but could lead to instability in a closed loop system.

6.2.4.4 P-T$_2$ Element

P behavior with 2nd order delay is referred to as *P-T$_2$*. The essential difference from P-T$_1$ elements lies in the oscillation ability of such systems. Typical examples of P-T$_2$ behavior are RLC circuits (see Fig. 6.41, see Sect. 3.6) and spring-mass oscillators (see Sect. 6.2.2.1). Depending on the damping, an aperiodic behavior such as that of the P-T$_1$ element, a periodic oscillation or even an unstable behavior can occur. However, the most common behavior is a slight overshoot or transient oscillation to a stable end value (Fig. 6.40).

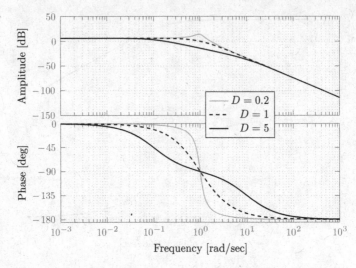

Fig. 6.40 Bode plot of a P-T$_2$ element ($K = 2$, $T = 1$, $D = 0.2$; 1; 5)

The functional relationship is $T^2\ddot{y}(t) + 2DT\dot{y}(t) + y(t) = Ku(t)$. The resulting transfer function is

$$G(s) = \frac{K}{1 + 2DTs + T^2s^2} \tag{6.30}$$

with D as the damping constant (see Sect. 6.2.3.5).

The amplitude and phase response in the Bode plot (see Fig. 6.40) result to

$$|G(j\omega)| = \frac{K}{\sqrt{(1 - T^2\omega^2)^2 + (2DT\omega)^2}} \tag{6.31}$$

$$\varphi(\omega) = -\arctan\left(\frac{2DT\omega}{1 - T^2\omega^2}\right) \tag{6.32}$$

The properties of the P-T$_2$ element depend on its damping. In physics, *damping* describes the transformation of the energy of a movement, an oscillation or a wave into another form of energy (see Sect. 6.2.3.5).

The characteristic equation follows from the denominator of the transfer function.

$$T^2s^2 + 2DTs + 1 = 0$$

The resulting poles of the transfer function are

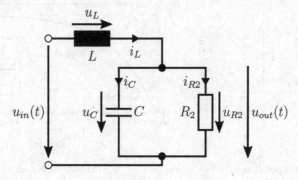

Fig. 6.41 Circuit implementation of a P-T$_2$ element

$$s_{1,2} = \frac{-2DT \pm \sqrt{4D^2T^2 - 4T^2}}{2T^2}$$

$$= \frac{1}{T}\left(-D \pm \sqrt{D^2 - 1}\right)$$

$$= -\omega_0 D \pm \omega_0\sqrt{D^2 - 1}$$

with D: *damping* and $\omega_0 = \frac{1}{T}$: *Natural frequency* of the undamped system ($D = 0$).

By interpreting the P-T$_2$ element as a series connection of two P-T$_1$ elements, the characteristic function can also be expressed as

$$T_1 \cdot T_2 \cdot s^2 + (T_1 + T_2) \cdot s + 1 = 0$$

resulting in the two time constants

$$T_1 = \frac{1}{s_1} \quad \text{and} \quad T_2 = \frac{1}{s_2}$$

This results in different properties according to the different pole positions.

For $D > 1$ two different negative real poles result (see Fig. 6.42)

$$s_1 = -\omega_0 D - \omega_0\sqrt{D^2 - 1}, \qquad s_2 = -\omega_0 D + \omega_0\sqrt{D^2 - 1}$$

The step response in the time domain results to

$$h(t) = K\left(1 - \frac{T_1}{T_1 - T_2}e^{-\frac{t}{T_1}} + \frac{T_2}{T_1 - T_2}e^{-\frac{t}{T_2}}\right)$$

The P-T$_2$ element therefore

- is not able to oscillate,
- leaves a residual steady-state error,
- is stable.

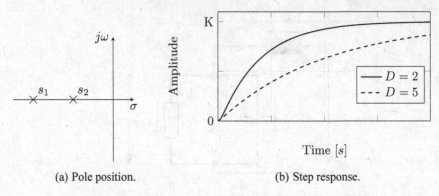

(a) Pole position.   (b) Step response.

Fig. 6.42 Pole-zero plot and step response of a P-T$_2$ element at a damping of $D > 1$

For $D = 1$ a double pole occurs (see Fig. 6.43). $s_{1,2} = -\omega_0 D$. The denominator for the transfer function becomes a binomial and this gives

$$G(s) = \frac{K}{(1 + Ts)^2}$$

Due to the double pole, T_1 and T_2 are equal and a linear term is created in the step response.

$$h(t) = K \left(1 - e^{-\frac{t}{T}} - \frac{t}{T} e^{-\frac{t}{T}} \right)$$

The system therefore

- is not able to oscillate,
- does not (necessarily) eliminate a steady-state error,
- is stable.

For $0 < D < 1$, a conjugated complex pole pair results

$$s_{1,2} = -\omega_0 D \pm j\omega_0 \sqrt{1 - D^2}$$

with a negative real part $\omega_0 D$ (see Fig. 6.44). The poles are located on the left side of the s plane on a circle with the radius ω_0. The angle ϑ is a measure of the damping D.

$$\cos \vartheta = D$$

The step response in the time domain results to

$$h(t) = K \left(1 - e^{-D\omega_0 t} \left[\cos\left(\sqrt{1 - D^2}\omega_0 t\right) + \frac{D}{\sqrt{1 - D^2}} \sin\left(\sqrt{1 - D^2}\omega_0 t\right) \right] \right)$$

The system therefore

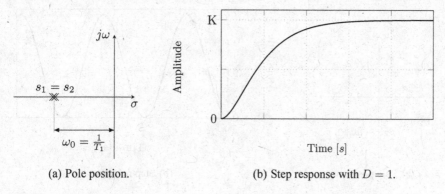

(a) Pole position.   (b) Step response with $D = 1$.

Fig. 6.43 Position of the poles and step response of a P-T$_2$ element with $D = 1$

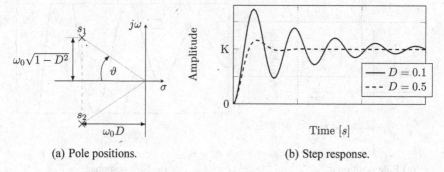

(a) Pole positions.   (b) Step response.

Fig. 6.44 Position of the poles and step response of a P-T$_2$ element with the damping $0 < D < 1$

- is able to oscillate,
- leaves a residual steady-state error,
- is stable.

For $D = 0$, the system is undamped and has the poles

$$s_{1,2} = \pm j\omega_0$$

(see Fig. 6.45). The case $D = 0$ is regarded as a marginal case of a stable P-T$_2$ element. Here the conjugated complex poles lie on the ordinate of the s-plane. A harmonic oscillation is obtained for the damping $D = 0$. The step response results to

$$h(t) = \frac{K}{T} \cdot sin\left(\frac{t}{T}\right)$$

The system therefore

- is able to oscillate,
- leaves a residual steady-state error,
- is marginally stable.

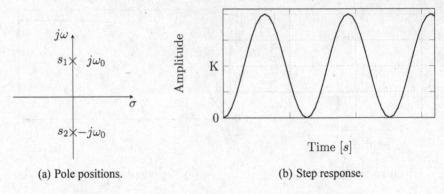

(a) Pole positions.   (b) Step response.

Fig. 6.45 Position of the poles and step response of a P-T$_2$ element with $D = 0$

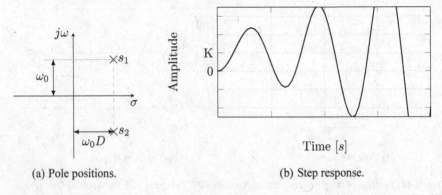

(a) Pole positions.   (b) Step response.

Fig. 6.46 Position of the poles and step response of a P-T$_2$ element with the damping $-1 < D < 0$

For $D < 0$, the system becomes unstable, its amplitude increases (see Fig. 6.46). The resulting step response for $-1 < D < 0$ corresponds to the step response for $0 < D < 1$, although the negative D leads to an increasing amplitude. For $D \leq -1$, the step response results to

$$h(t) = K \left(1 - \frac{T_1}{T_1 - T_2} e^{\frac{t}{T_1}} + \frac{T_2}{T_1 - T_2} e^{\frac{t}{T_2}} \right)$$

It has the following characteristics:

- able to oscillate (for $-1 < D < 0$, not below anymore, see Fig. 6.47),
- leaves a residual steady-state error,
- unstable.

In summary, it can be said for the damping of the P-T$_2$ element that the following applies to the poles of $G(s) = \frac{K}{T^2s^2 + 2DTs + 1}$ $s_{1,2} = \frac{1}{T} \left(-D \pm \sqrt{D^2 - 1} \right)$:

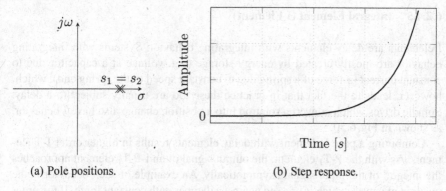

(a) Pole positions.

(b) Step response.

Fig. 6.47 Position of the poles and step response of a P-T$_2$ element with $D \leq -1$

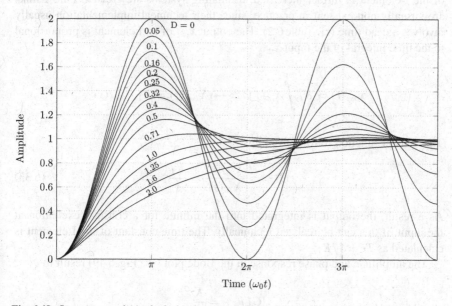

Fig. 6.48 Step response $h(t)$ of a 2nd order system able to oscillate for various dampings D

- For $D > 1$ negative real poles $\Rightarrow$ overdamped case,
- for $D = 1$ double pole $\Rightarrow$ overdamped case,
- for $D < 1$ conjugated complex poles $\Rightarrow$ damped oscillation,
- for $D = 0$ purely imaginary poles $\Rightarrow$ undamped oscillation,
- for $D < 0$ positive real part $\Rightarrow$ oscillation with increasing amplitude.

The rise and settling times of the control also depend on the damping. The greater the damping D, the slower $h(t)$ reaches the end value for the first time, on the other hand the settling time is shorter (see Fig. 6.48).

6.2.4.5   Integral Element (I Element)

I elements are delay elements with integrating behavior. Systems with integrating behavior are mostly created by energy storage (e.g. voltage at a capacitor due to constant current or force of a spring due to constant speed of the spring end), which, however, leads to the fact that in practice these too are usually subject to a delay. Spindle drives which convert a rotation into a position change also have I behavior, as shown in Fig. 6.50.

Combining a pure I element with delay elements results in higher order I-T elements. As with the P-T$_1$ element, the output signal of an I-P-T$_1$ element approaches the integral of the input signal asymptotically. An example of such behavior is the position of a mass which is pressed against a damper with constant force. If the order of the I-element is further increased, oscillating systems are created. Pure I links are virtually non-existent in practice, since their technical implementation usually involves a dead time (cf. Table 6.2). The output x_{out} of an I element is proportional to the time integral of the input x_{in}.

$$x_{out}(t) = K_I \int\limits_{0}^{t} x_{in}(\tau)d\tau \mathrel{\multimap\bullet} s\, X_{out}(s) = K_I \cdot X_{in}(s)$$

The transfer function follows from this

$$G(s) = \frac{X_{out}(s)}{X_{in}(s)} = \frac{K_I}{s} \tag{6.33}$$

As a result, the output is integrated into the infinite for a constant excitation at the input, if this can be realized technically. The time constant of the I element is calculated as $T_I = 1/K_I$.

The amplitude and phase response in the Bode plot (see Fig. 6.49) result to

$$|G(j\omega)| = \frac{K}{\omega}$$

$$\varphi(\omega) = -\frac{\pi}{2}$$

Properties of the I element:

- I elements are not capable of oscillating individually, but multiple elements are.
- I elements do not eliminate a residual steady-state error in an open control loop, but in a feedback circuit they do. *Only* if a system also contains I behavior, it has no residual steady-state error.
- I elements are inherently unstable or marginally stable and can cause instabilities in the system as part of the system or controller.

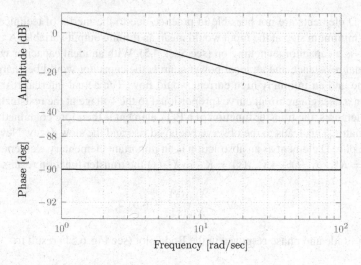

Fig. 6.49 Bode plot of an I element ($K = 2$)

Fig. 6.50 Example: track of a spindle drive

Table 6.2 Comparison of different I elements

Transfer function	I element	I/P-T_1 element	I/P-T_2 element
$G(s)$	$\frac{K_I}{s}$	$\frac{K_I}{s(Ts+1)}$	$\frac{K_I}{s(s^2T^2+2DTs+1)}$
$h(t)$	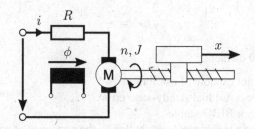		

- The I controller takes the integration time T_I *longer* for the control process (Fig. 6.50).

6.2.4.6 Differential Element (D Element)

D elements are delay elements with differentiating behavior. This means that the output x_{out} of a system is proportional to the time derivative of the input x_{in}.

Pure D elements are not possible in practice due to the inertia of technical systems. A minimum step at the input would mean an infinite output variable. A typical example is a capacitor switching on (see Sect. 3.5). With an ideal capacitor without an internal resistance and an ideal power source, the capacitor would be charged in zero time and an infinitely high current would flow. The actual internal resistance results in a decaying current curve (proportional to the voltage at the resistor). With real D elements, the transfer function of a P-T$_1$ element is therefore contained in the denominator. This leads to the observed delayed rise and the slow decay. Nevertheless, the pure D element as an abstraction is an important elementary element. From $x_{out}(t) = K_D \cdot x_{in}(t) \circ\!\!-\!\!\bullet X_{out}(s) = K_D \cdot s X_{in}(s)$ the transfer function results to

$$G(s) = \frac{X_{out}(s)}{X_{in}(s)} = K_D \cdot s \tag{6.34}$$

The amplitude and phase response in the Bode plot (see Fig. 6.51) result to

$$|G(j\omega)| = K\omega$$

$$\varphi(\omega) = \frac{\pi}{2}$$

Properties of the D element:

- D elements are not capable of oscillating.
- D elements leave a residual steady-state error.
- D elements are not BIBO-stable.
- D elements make a significant contribution to the dynamics of a system.
- Ideal D elements cannot be realized individually. However, they are often approximated.

6.2.4.7 PI Element

The PI element is a composite element that is mainly used as a regulator. At the same time, it represents a real variant of the I-element. When used as a controller, it tries to compensate for the disadvantages of the two individual controllers by combining P and I controllers (P controller: no stationary accuracy, I controller: slow). From the temporal behavior

$$x_{out}(t) = K_P \cdot \left(e(t) + \frac{1}{T_I} \int_0^t x_{in}(\tau) d\tau \right)$$

the transfer function results to

$$G_{PI}(s) = K_P \frac{1 + T_I s}{T_I s}$$

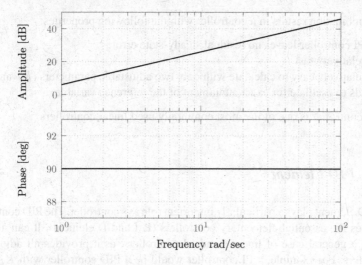

Fig. 6.51 Bode plot of a D element ($K = 2$)

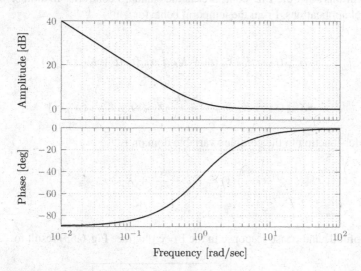

Fig. 6.52 Bode plot of a PI element ($K_P = T_I = 1$)

The amplitude and phase response in the Bode plot (see Fig. 6.52) result to

$$|G(j\omega)| = K_P \cdot \sqrt{1 + \frac{1}{T_I^2 \omega^2}}$$

$$\varphi(\omega) = -\arctan\left(\frac{1}{T_I \omega}\right)$$

This combination results in a controller with the following properties:

- The PI controller leaves no residual steady-state error.
- It is relatively fast.
- It is relatively easy to calculate with only two adjustable parameters (K_P and T_I).
- It tends to oscillate for exact attainment of the reference variable.

The PI controller is one of the most commonly used linear controllers.

6.2.5 PID Element

The PID element plays a particularly important role as a controller. The PID controller combines the essential elementary controllers (P, I and D element). It can also be seen as a general case of linear standard controllers, as it provides all adjustable parameters. For example, a PI controller would be a PID controller with $K_D = 0$. Viewed in this way, the PID controller is the standard controller in almost 90% of industrial applications. From the temporal behavior

$$x_{out}(t) = K_P x_{in}(t) + K_I \int\limits_0^t x_{in}(t)dt + K_D \cdot \dot{x}_{in}(t)$$

$$\circ\!\!-\!\!\bullet X_{out}(s) = K_P X_{in}(s) + \frac{K_I}{s} X_{in}(s) + K_D \cdot s X_{in}(s)$$

its transfer function in the complex variable domain is

$$G_{PID}(s) = K_P + \frac{K_I}{s} + K_D \cdot s$$

The amplitude and phase response in the Bode plot (see Fig. 6.53) result to

$$|G(j\omega)| = \sqrt{K_P^2 + \left(\frac{K_I}{\omega} + K_D\omega\right)^2}$$

$$\varphi(\omega) = -\arctan\left(\frac{K_D\omega + \frac{K_I}{\omega}}{K_P}\right)$$

The PID controller has the following features:

- The PID controller leaves no residual steady-state error.
- It is fast (because of P and D elements).

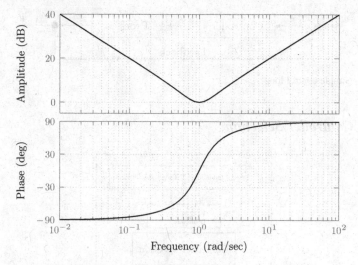

Fig. 6.53 Bode plot of a PID element ($K_P = K_I = K_D = 1$)

- It also tends to oscillate in order to achieve the exact reference variable, but this can be damped more easily by the D component than in the PI controller.

Finally, a short overview of the system elements presented so far is provided below. This overview is by no means exhaustive, but covers an important part of the models used in control engineering. For comparison purposes, in addition to the transfer function in the complex variable domain, its equivalent in the time domain is also shown here.

System	Transfer function $G(s)$	Step response
P	K_P	$K_P \cdot \sigma(t)$
T_t	$K_P \cdot e^{-sT_t}$	$\sigma(t - T_t)$
$P - T_1$	$\frac{K_P}{1+sT_1}$	$K_P \cdot \left(1 - e^{-\frac{t}{T_1}}\right) \cdot \sigma(t)$
$P - T_2 \ (0 < D < 1)$	$\frac{K_P}{1+2 \cdot D \cdot sT_0 + (sT_0)^2}$	$K_P \left(1 - e^{-D\omega_0 t}\left[\cos\left(\sqrt{1-D^2}\omega_0 t\right)\right.\right.$ $\left.\left. +\frac{D}{\sqrt{1-D^2}}\sin\left(\sqrt{1-D^2}\omega_0 t\right)\right]\right) \cdot \sigma(t)$ with $\omega_0 = \frac{1}{T_0}$
I	$\frac{1}{sT_I}$	$K_I \cdot t \cdot \sigma(t) = \frac{t}{T_I} \cdot \sigma(t)$ with $T_I = \frac{1}{K_I}$ as the integration time
D	sT_V with $T_V = K_D$	$K_D \cdot \delta(t)$
PI	$K_P \frac{1+T_I s}{T_I s}$	$(K_P + \frac{t}{T_I}) \cdot \sigma(t)$
PID	$K_P + \frac{K_I}{s} + K_D s$	$(K_P + t \cdot K_I) \cdot \sigma(t) + K_D \cdot \delta(t)$

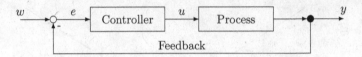

Fig. 6.54 Simple control loop

6.2.6 Controller Selection

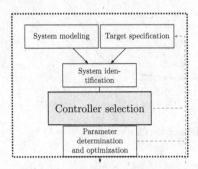

Following on from the consideration of modeling and identification of the system, the actual control will now be examined. The first step is to select a suitable controller. In doing so, one must first take into account which of the objectives the system already fulfils in itself (which results from modeling and identification, see Sect. 6.2.2) and which are still missing (which results from the defined target parameters, see Sect. 6.2.1). To achieve this, the known properties of the standard elements are used, at least for simple linear controllers. These are added to the system accordingly to achieve the desired behavior (see Fig. 6.54).

It should be noted that this procedure not only complements the properties of the new elements, but can also possibly change previous behavior by overlaying "old" and "new" elements. This is desirable in part (inappropriate characteristics of the system can be eliminated), but it can also lead to problems. (e.g. the new system has stationary accuracy, but unlike the old one, it can oscillate) A re-evaluation of the overall system using the methods described above is therefore essential. In order to better assess the properties of possible controllers, some examples will now be compared.

Let us consider, for example, the selection of a controller for a P-T$_1$ system:

The *modeling* is limited to finding the transfer function of the system, which by definition is a P-T$_1$ element:

$$G_S(s) = \frac{K_{Sys}}{1 + sT_1}$$

Identification: The P-T$_1$ element is known not to eliminate a steady-state error in itself because there is no I-component present. The amplification is constant, with increasing angular frequency the phase shift is aiming towards $-90°$.

Controller selection To ensure the stationary accuracy of the overall system, the controller must have an I component (since only an I component can provide stationary accuracy). However, since a pure I controller is very slow and the phase shift of $-90°$ further endangers stability, a PI controller is preferred. This is not only faster, but can also shift the stability limit by means of proportional amplification. If it does not prove to be sufficient, a D component can be added.

The transfer function of the PI controller is

$$G_R(s) = \frac{K_P(1 + sT_n)}{sT_n}$$

Using Eq. (6.9), the transfer function of the feedback system results in

$$G_w(s) = \frac{G_S(s) \cdot G_R(s)}{1 + G_S(s) \cdot G_R(s)} = \frac{1}{\frac{(1+sT_1)sT_n}{K_P K_{Sys}(1+sT_n)} + 1}$$

or by conversion to

$$G_w(s) = \frac{K_P K_{Sys}(1 + sT_n)}{(1 + sT_1)sT_n + K_P K_{Sys}(1 + sT_n)}$$

This quite extensive formula can be simplified by means of a trick. When considering the parameters, it is noticeable that two different time constants occur here. Since the time constants of the controller are adjustable parameters, they can be selected as equal to the time constants of the system.
Control parameter $T_n = T_1$:

$$G_w(s) = \frac{K_P K_{Sys}(1 + sT_1)}{(1 + sT_1)(K_P K_{Sys} + sT_1)} = \frac{K_P K_{Sys}}{K_P K_{Sys} + sT_1}$$

The resulting system should now be re-examined to ensure that the desired objectives are achieved.
Steady-state behavior:

$$\lim_{t \to \infty} y(t) = \lim_{s \to 0} s \cdot W(s) \cdot G_w(s)$$
$$= G_w(0) \cdot \lim_{s \to 0} s \cdot W(s)$$
$$= 1 \cdot \lim_{t \to \infty} w(t)$$
$$= w_0$$

This results in the control difference (see Sect. 6.2.1)

$$e(\infty) = w_0 - y(\infty) = 0$$

Thus the system does not leave a residual steady-state error. In order to be able to make statements about the stability of the system, concrete values must be available for the system. For the sake of simplicity, the example is based on $K_{sys} = 1$ and $T_1 = 1$ s. (Note: Due to the occurrence of "second" and the "s" in the Laplace transformation, caution is required when calculating! Usually such units are ignored uring the calculation.) This results in the transfer function of the overall system

$$G_w(s) = \frac{K_P}{K_P + s}$$

For a stability analysis with a pole-zero plot, poles and zeros of s are calculated. Poles: s: $p_1 = -K_P$; there are no zeros. So as long as $K_P >= 0$ is valid, the system is stable.

In order to now find a suitable value for K_P, further considerations must be included, which will be presented in the next section on parameter definition.

6.2.7 Parameter Determination

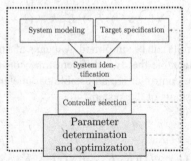

Once a controller has been found that appears to meet the relevant objectives, it must be adapted to the specific system. For this purpose, the parameters of the controller must be set in such a way that the original objectives are fulfilled as far as possible. Normally, this means that the controller should be stable and capable of eliminating the steady-state error and react as fast as possible with as little overshoot as possible to changes in the input variable. At the same time it should be highly robust against disturbances.

The properties of the overall system, including the controller, are examined mathematically to achieve this objective, as shown in the previous example of controller selection. This means that the overall transfer function for the closed control loop including the controller is set up on the basis of still undefined controller parameters. The following are calculated:

- Pole and zero points for stability considerations,

- Step response for overshoots and speed,
- Phase reserve and/or amplitude reserve (Bode plot/Nyquist plot) for robustness.

Using the formulas found in this way, dependencies of the target parameters on the controller parameters $(K_I, K_P, T_D,$ etc.) can be identified. These are adapted to requirements by means of appropriate mathematical methods.

6.2.7.1 In summary, the classic procedure for controller design is as follow

Now that the classic, model-based design of linear controllers has been considered, here is a brief summary of the procedure presented:

1. Specification of the target parameters
2. Modeling of the system (math. description of technical processes)

   - Input/output relationships and status representations for subsystems as well as linkage rule
   - Theory and algorithms for the experimental determination of system models and their parameters
   - Theory and algorithms for model simplification

3. Analysis/identification (recording of system properties)

   - Stability of the closed control loop including robustness analysis
   - Behavior of the control loop with regard to dynamics, steady-state error, interference suppression and robustness against parameter fluctuations

4. Synthesis (controller selection and parameter determination)

   - Design methods to meet given specifications
   - Design methods for signal processing and state estimation
   - Design methods for the optimization of control loops

6.2.7.2 Control of a DC Motor for an Autonomous Forklift Truck with Differential Drive

In this example a speed control is developed for the autonomous forklift truck, for which we assume a differential drive mechanism by means of DC motors (see Sect. 9.3.1). Thus, the controlled variable is the rotational speed of the motors. Of course, the system should be stable and leave no residual steady-state error in operation. Overshoots should be avoided or kept as small as possible. We can use the armature voltage of the motors as the manipulated variable (see Fig. 6.55). The target parameters of the system are thus already fixed.

First, the transfer function of the motors must be set up so that the controller can be designed appropriately. This corresponds to the modeling and identification of the

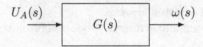

$$U_A(s) \longrightarrow \boxed{\quad G(s) \quad} \longrightarrow \omega(s)$$

Fig. 6.55 Black-box of a DC motor

system. For this purpose we use the electrical model presented in Sect. 9.3.1. Only the armature voltage is available to us for controlling the DC motors. It can be used to influence the armature current and thus the drive torque T_A, which is described by Eq. (9.3). This in turn influences the rotational speed. The result is the general transfer function in Eq. (6.35).

$$G(s) = \frac{\omega(s)}{U_A(s)} \tag{6.35}$$

The DC motor is an LTI system (see Sect. 5.2.6), since all differential equations are linear. Therefore, the Laplace transformation can be applied. In order to establish a closed transfer function, the required electrical equations of the DC motor are transferred from the time domain to the Laplace domain.

The transformation of Eq. (9.6) solved with I_A gives Eq. (6.36).

$$I_A = \frac{\dfrac{1}{R_A}}{1 + \dfrac{L_A}{R_A} \cdot s} \cdot (U_A(s) - U_{ind}(s)) \tag{6.36}$$

The transformation of Eq. (9.9) solved with ω gives Eq. (6.37).

$$\omega(s) = \frac{1}{J \cdot s} \cdot T_B(s) \tag{6.37}$$

From Eqs. (9.3) and (9.4), the following also results.

$$T_A(s) = K_F \cdot I_A(s) \tag{6.38}$$

and

$$U_{ind}(s) = K_F \cdot \omega(s) \tag{6.39}$$

The block diagram in Fig. 6.56 can now be put together from these relationships. This results in the control difference $U_R = U_A - U_{ind}$.

The load torque T_B cannot be used in the calculation because it is still unknown as a disturbance variable at the time of calculation and cannot be used as a constant. Thus this disturbance is assumed to be $T_L = 0$ and the result is

$$\omega(s) = \frac{1}{J \cdot s} \cdot T_A(s) \tag{6.40}$$

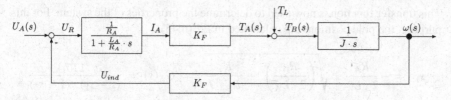

Fig. 6.56 Block diagram of a DC motor

In order to summarize the transfer function, the above equations are merged.

$$
\begin{aligned}
U_R(s) &= U_A(s) - U_{ind}(s) = U_A(s) - \omega(s) \cdot K_F \\
&= U_A(s) - \frac{T_A(s)}{J \cdot s} \cdot K_F \\
&= U_A(s) - \frac{I_A(s) \cdot K_F}{J \cdot s} \cdot K_F \\
&= U_A(s) - \frac{K_F^2}{J \cdot s} \cdot \frac{\frac{1}{R_A}}{1 + \frac{L_A}{R_A} \cdot s} \cdot U_R(s)
\end{aligned}
\tag{6.41}
$$

Further conversion gives

$$
U_R(s) = \frac{U_A(s)}{1 + \frac{K_F^2}{J \cdot s} \cdot \dfrac{\frac{1}{R_A}}{1 + \frac{L_A}{R_A} \cdot s}}
\tag{6.42}
$$

Inserting the formula (6.39) gives

$$
U_A(s) \cdot \left[1 - \frac{1}{1 + \frac{K_F^2}{J \cdot s} \cdot \dfrac{\frac{1}{R_A}}{1 + \frac{L_A}{R_A} \cdot s}} \right] = U_{ind}(s) = K_F \cdot \omega(s)
\tag{6.43}
$$

This results in the transfer function

$$
G(s) = \frac{\omega(s)}{U_A(s)} = \frac{\frac{K_F}{R_A}}{J \cdot s \cdot \left(1 + \frac{L_A}{R_A} \cdot s \right) + \frac{K_F^2}{R_A}}
\tag{6.44}
$$

$$
\iff G(s) = \frac{K_F}{L_A \cdot J \cdot s^2 + R_A \cdot J \cdot s + K_F^2} = \frac{\frac{K_F}{L_A \cdot J}}{s^2 + \frac{R_A}{L_A} \cdot s + \frac{K_F^2}{L_A \cdot J}}
$$

This transfer function is now used to determine the properties of the system. For this purpose, the pole positions are calculated in Eq. (6.45).

$$s_{1,2} = -\frac{R_A}{2 \cdot L_A} \pm \sqrt{\left(\frac{R_A}{2 \cdot L_A}\right)^2 - \frac{K_F^2}{L_A \cdot J}} \implies G_S(s) := \frac{\frac{K_F}{L_A \cdot J}}{(s - s_1) \cdot (s - s_2)}$$

(6.45)

The DC motors used in this example provide a maximum armature voltage of 24 V. They are controlled by a microcontroller (see Sect. 10.4) which has two 12-bit PWM units (see Sect. 7.7.3). Thus it is possible to control the armature voltage from 0 to 24 V in $2^{12} = 4096$ steps. The proportional transfer function for this translation is described in Eq. (6.46).

$$G_{PWM}(s) = P_{PWM} = \frac{24}{4096}$$

(6.46)

The angular velocity ω of the motor shaft is measured by an integrated encoder, which generates 512 pulses for each revolution of the shaft in two channels, which have a phase shift of $90°$. Thus 2048 step values for one revolution are generated. The ratio between encoder resolution and motor revolutions is calculated by Eq. (6.47).

$$2048 \frac{\text{encoder step values}}{s} \longleftrightarrow \omega = \frac{2 \cdot \Pi}{s}$$

(6.47)

The values generated by the encoder are sent via the board to the decoder of the microcontroller. This results in a proportional delay between the measurement at the motor ω and the evaluation at the microcontroller enc. The controller has a clock rate of about 1 ms, therefore specification of ω in kHz is indicated. The transfer function is described in Eq. (6.49).

$$enc = \frac{2048}{2 \cdot \Pi} \cdot \omega$$

(6.48)

$$G_{enc}(s) = P_{enc} = \frac{2048}{2 \cdot \Pi} \cdot 10^{-3}$$

(6.49)

Since the complete system has two pole positions, it can become capable of eliminating the steady-state error, but does not have to. The controller should therefore have integrating behavior. The speed of the movement should remain as constant as possible, but one PI controller seems sufficient for the speed requirements. The transfer function of the PI controller (Eq. (6.50)) depends on the proportionality factor K_C and the time constant for the integration T_I.

$$G_{PI}(s) = K_C \cdot \frac{T_I \cdot s + 1}{T_I \cdot s}$$

(6.50)

With this knowledge the block diagram of the control in Fig. 6.57 can be derived. enc_{des} is the desired number of encoder values per cycle as system input and enc_{mes} is the output value describing the number of encoder values in relation to the angular

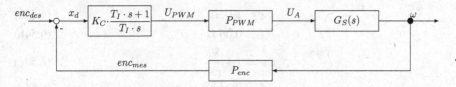

Fig. 6.57 Block diagram of the closed control loop including the PI controller and the DC motor

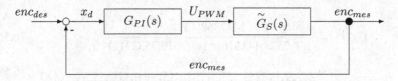

Fig. 6.58 Block diagram of the closed control loop with negative feedback

velocity ω of the motor shaft. x_d is the control difference and u_{PWM} is the PWM value at the controller output.

The parameters K_C and T_I can be determined using a Bode plot. For this purpose, the proportional variables P_{PWM} and P_{enc} are combined to $\tilde{G}_S(s)$. The modified block diagram can be seen in Fig. 6.58. The system output now corresponds to the shaft rotation in encoder measurements per clock pulse. The adjusted controlled system is described by Eq. (6.51). The open-loop Eq. (6.52) and the final closed-loop Eq. (6.53) can be derived from this equation.

$$\tilde{G}_S(s) = \frac{P_{PWM} \cdot P_{enc} \cdot \frac{K_F}{L_A \cdot J}}{(s - s_1) \cdot (s - s_2)} \tag{6.51}$$

$$F_{out}(s) = G_{PI}(s) \cdot \tilde{G}_S(s) = K_C \cdot \frac{T_I \cdot s + 1}{T_I \cdot s} \cdot \frac{P_{PWM} \cdot P_{enc} \cdot \frac{K_F}{L_A \cdot J}}{(s - s_1) \cdot (s - s_2)} \tag{6.52}$$

$$G_{tot} = \frac{enc_{mes}(s)}{enc_{des}(s)} = \frac{F_{out}(s)}{1 + F_{out}(s)} \tag{6.53}$$

To determine the parameters for the PI controller, the pole positions are calculated with Eq. (6.45). The electrical and mechanical parameters of the DC motors used are taken from the data sheets (see Eqs. (6.54)) and inserted into the Eq. (6.51). This gives Eq. (6.55).

$$L_A = 19 \cdot 10^{-6} \text{ H} \tag{6.54a}$$

$$R_A = 0.06 \, \Omega \tag{6.54b}$$

$$K_F = 1.1 \, \frac{\text{Vs}}{\text{rad}} \tag{6.54c}$$

$$J = 0.04 \text{ kgm}^2 \tag{6.54d}$$

$$\tilde{G}_S(s) \approx \frac{1447368.421}{(s + 2528.141516) \cdot (s + 629.7532212)} \tag{6.55a}$$

$$\approx \frac{0.90909}{(0.000395547 \cdot s + 1) \cdot (0.001587924 \cdot s + 1)} \tag{6.55b}$$

For a sufficiently high control speed, T_I is chosen so that the dominant time constant (in the denominator) can be compensated. In this example, $T_I \approx 0.001587924$ is used and this gives Eq. (6.56).

$$F_{out}(s) \approx K_C \cdot \frac{0.001587924 \cdot s + 1}{0.001587924 \cdot s} \cdot \frac{0.90909}{(0.000395547 \cdot s + 1) \cdot (0.001587924 \cdot s + 1)} \tag{6.56a}$$

$$\approx \frac{572.5 \cdot K_C}{0.000395547 \cdot s^2 + s} \tag{6.56b}$$

In order to achieve a high stability at the appropriate control speed, K_C is selected in such a way that the phase margin corresponds to $\phi_R \approx 60°$ (distance of the phase curve $F_{out}(j\omega)$ from $-180°$ at the zero point). The evaluation of the Bode plot in Fig. 6.59 of the control transfer function gives $K_C \approx 2.9$. A curve of the system controlled in this way under changing load and speed is shown in Fig. 6.60.

6.2.7.3   Control of an Inverse Pendulum

The inverse pendulum is a rod which is mounted on a carriage with a freely movable hinge joint. The carriage moves in a straight line and should raise and balance the rod by its movement. For this example, a simplified model with a one-dimensional linear drive is used (see also Sect. 6.3.2).

The identifiers in Fig. 6.61 describe the following variables:

M = mass of the carriage, m = mass of the rod, l = length of the rod, b = friction of the carriage on the ground, N = horizontal forces between carriage and rod, P = vertical forces between carriage and rod, I = moment of inertia of the rod, θ = angle between rod and vertical, x = position of the carriage, $\dot{x}$ = speed of the carriage and $\ddot{x}$ = acceleration of the carriage.

In order to describe the balance of forces of the carriage mathematically, the inertia forces of the carriage $M\ddot{x}$, the frictional forces $b\dot{x}$ and the force applied horizontally by the rod N are equated with the driving force F of the carriage.

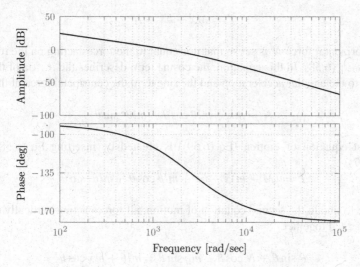

Fig. 6.59 Bode plot of the open control loop. For $K_C \approx 2.9$ an inclination angle of $\approx 60°$ can be seen

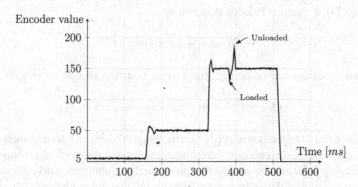

Fig. 6.60 Reaction of the speed controller to changes in the input and to disturbances

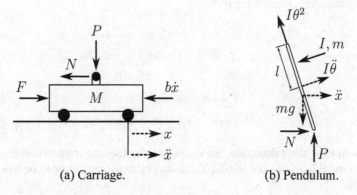

(a) Carriage.   (b) Pendulum.

Fig. 6.61 Forces and states of the inverse pendulum

$$F = M\ddot{x} + b\dot{x} + N \tag{6.57}$$

If the horizontal force N is set so that it is equal to the forces acting on the rod, the result is Eq. (6.58). In this equation, the cosine term describes the inertia of the rod relative to the angular acceleration and the sine term the centripetal force of the rod.

$$N = m\ddot{x} + ml\ddot{\theta}\cos\theta - ml\dot{\theta}^2\sin\theta \tag{6.58}$$

The first equation of motion (Eq. (6.59)) is created by inserting Eq. (6.58) into Eq. (6.57).

$$F = (M + m)\ddot{x} + b\dot{x} + ml\ddot{\theta}\cos\theta - ml\dot{\theta}^2\sin\theta \tag{6.59}$$

In order to obtain the second equation of motion, all forces acting vertically on the rod are added together.

$$P\sin\theta + N\cos\theta - mg\sin\theta = ml\ddot{\theta} + m\ddot{x}\cos\theta \tag{6.60}$$

In order to replace the (unknown) horizontal and vertical forces between carriage and rod (P, N), a moment balance is drawn up.

$$-Pl\sin\theta - Nl\cos\theta = I\ddot{\theta} \tag{6.61}$$

If the two equations are combined, the second equation of motion (Eq. (6.62)) is obtained.

$$(I + ml^2)\ddot{\theta} + mgl\sin\theta = -ml\ddot{x}\cos\theta \tag{6.62}$$

In order to be able to analyze the two equations of motion with formal methods of control engineering, the equations are linearized around the vertical position ($\theta = \pi = operating\ point$) of the rod. In this procedure, the nonlinear curve around this point is approximated with a linear curve, assuming that the rod moves only slightly out of the vertical position. This results in $\theta = \pi + \phi$. This in turn results in the following simplifications in Eq. (6.63).

$$\cos\theta = -1 \tag{6.63a}$$

$$\sin\theta = -\phi \tag{6.63b}$$

$$\dot{\theta}^2 = 0 \tag{6.63c}$$

$$(I + ml^2)\ddot{\phi} - mgl\phi = ml\ddot{x} \tag{6.63d}$$

$$(M + m)\ddot{x} + b\dot{x} - ml\ddot{\phi} = u \tag{6.63e}$$

In order to ease the calculation, we again use the Laplace transformation, which results in the Eqs. (6.64) and (6.65). The initial values are assumed to be 0 here.

$$(I + ml^2)\Phi(s)s^2 - mglx\,\Phi(s) = mlX(s)s^2 \tag{6.64}$$

$$(M + m)X(s)s^2 + bX(s)s - ml\Phi(s)s^2 = U(s) \tag{6.65}$$

In order to obtain a relationship between the manipulated variable F and the output variable I, the Eq. (6.64) is solved according to $X(s)$ (Eq. (6.66)) and inserted into Eq. (6.65). This results in Eq. (6.67).

$$X(s) = \left[\frac{(I + ml^2)}{ml} - \frac{g}{s^2} \right] \Phi(s) \tag{6.66}$$

$$(M + m)\left[\frac{(I + ml^2)}{ml} - \frac{g}{s^2} \right] \Phi(s)s^2 + b \left[\frac{(I + ml^2)}{ml} - \frac{g}{s^2} \right] \Phi(s)s - ml\Phi(s)s^2 = U(s)$$

$$\tag{6.67}$$

After conversion, the transfer function (6.68) with $q = (M + m)(I + ml^2) - (ml)^2$ is obtained.

$$G_{Sys}(s) = \frac{\Phi(s)}{U(s)} = \frac{\frac{ml}{q}s}{s^3 + \frac{b(I+ml^2)}{q}s^2 - \frac{(M+m)mgl}{q}s - \frac{bmgl}{q}} \tag{6.68}$$

The vertical position of the pendulum is an unstable resting position, as any slight disturbance can cause the pendulum to topple over. In order to stabilize the pendulum in the vertical position, a controller is required. The controller must counteract the disturbance variable as quickly as possible and no control difference must remain, as otherwise the carriage will try to accelerate to infinity as compensation. A PID controller is therefore selected for stabilization.

In order to determine the parameters K_P, K_I and K_D of the controller, the transfer function of the closed loop must first be set up. The transfer function can be set up using the formula for the reference variable behavior, as can be seen in Eq. (6.69). The parameters for the transfer function are assumed for this example to be as follows.

$M = 2\,\text{kg}, m = 0.5\,\text{kg}, l = 0.3\,\text{m}, i = m \cdot L^2 = 0.045\,\text{kgm}^2, b = 0.1\,\frac{\text{N}}{\text{ms}}$ and $g = 9.8\,\frac{\text{m}}{\text{s}^2}$

In order to be able to solve equations more easily, transfer functions are described by a numerator term $NUM(s)$ and a denominator term $DEN(s)$ and only replaced by the corresponding functions in the last step.

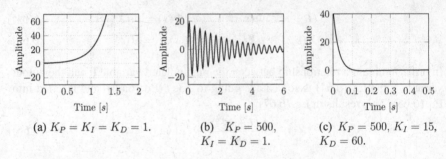

(a) $K_P = K_I = K_D = 1.$ (b) $K_P = 500,$ (c) $K_P = 500,\ K_I = 15,$
   $K_I = K_D = 1.$ $K_D = 60.$

Fig. 6.62 Impulse response of the system with the different controller parameters

$$G_R(s) = \frac{K_D \cdot s^2 + K_P \cdot s + K_I}{s} = \frac{p}{s} \tag{6.69a}$$

$$G_0(s) = G_R(s) \cdot G_{Sys}(s) = \frac{NUM_R(s)}{DEN_R(s)} \cdot \frac{NUM_{Sys}(s)}{DEN_{Sys}(s)} \tag{6.69b}$$

$$G_{tot}(s) = \frac{G_0(s)}{1 + G_0(s)} = \frac{NUM_R(s) \cdot NUM_{Sys}(s)}{DEN_R(s) \cdot DEN_{Sys}(s) + NUM_R(s) \cdot NUM_{Sys}(s)} \tag{6.69c}$$

$$= \frac{p \cdot (\frac{ml}{q}s)}{\left(s^3 + \frac{b(I+ml^2)}{q}s^2 - \frac{(M+m)mgl}{q}s - \frac{bmgl}{q}\right) \cdot s + p \cdot (\frac{ml}{q}s)} \tag{6.69d}$$

$$= \frac{(K_D \cdot s^2 + K_P \cdot s + K_I) \cdot 0.741s}{(s^3 + 0.044s^2 - 18.15s - 0.726) \cdot s + (K_D \cdot s^2 + K_P \cdot s + K_I) \cdot 0.741s} \tag{6.69e}$$

$$= \frac{0.7407s^3 + 0.7407s^2 + 0.7407s}{s^4 + 0.7852s^3 - 17.41s^2 + 0.01481s} \tag{6.69f}$$

The impulse response of the closed control loop can be calculated by transformation back into the time domain. In the example of the inverse pendulum, to get the impulse response means, a short impact (Dirac impulse) is given on the carriage. As the system response, the pendulum is brought out of its resting position by the impact. In order to determine the parameters of the controller, the impulse response is first calculated for $K_P = K_I = K_D = 1$. By inserting these controller parameters into Eq. (6.69e) the transfer function in Eq. (6.69f) is obtained.

In Fig. 6.62a, the still unstable behavior of the system is evident. As the linearized system is studied, the angle between the vertical and the pendulum increases to infinity. Now the proportional part of the controller K_P is increased until a decaying oscillation around the vertical position occurs. The oscillation for $K_P = 500$ is shown in Fig. 6.62b. Once the value for K_P has been determined, the differential part K_D is increased until the decaying oscillation becomes a single overshoot. In order to reach the stationary final value faster, K_I is increased slightly. Figure 6.62c shows

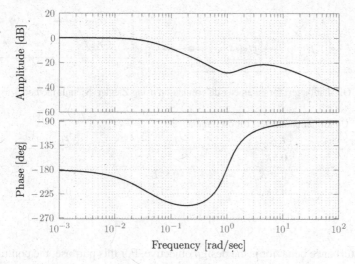

Fig. 6.63 Bode plot with the controller parameters $K_P = K_I = K_D = 1$

the impulse response of the set controller. The impulse seems to come from infinity, since an ideal D component is assumed for the controller.

Another way to set the controller parameters is to use the frequency response design method. As in the previous example, the Bode plot is used here. Again, a phase margin ϕ_R of approx. 60° should be maintained in order to be robust with regard to interference. This method is particularly suitable for working out the stability limit or robustness limit of the system.

Figure 6.63 shows the Bode plot for $K_P = K_I = K_D = 1$. There is no phase margin at gain crossover frequency of $\omega_0 = 8.11 \cdot 10^{-3}$ rad/s. The closed control loop is unstable. If the numerator term of the controller is extended by the controller gain K_R, the course of the amplitude response can be shifted up or down. The phase response remains unaffected by this. If the gain is increased in the example of the inverse pendulum, the gain crossover frequency moves further to the right. If the gain crossover frequency is above 0.98 rad/s, the control loop is stable. This corresponds to a gain of approx. 26.

6.2.8 Heuristic Method According to Ziegler–Nichols

In addition to the classical approach to controller design, various empirical methods have been developed over the past decades. These offer great savings in time and effort, especially for relatively simple systems or very loose requirements. The Ziegler and Nichols method offers a simple and fast method of this sort for designing a controller which is sufficient in many cases. Strictly speaking, these are two methods based on vibration testing or step response measurement. With this method,

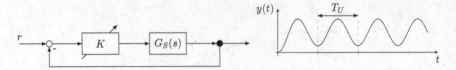

Fig. 6.64 Oscillation test for parameter determination using Ziegler–Nichols method 1

Table 6.3 Controller parameters for oscillation tests

Controller	K_P	T_N	T_D
P	$0.5K_u$	–	–
PI	$0.45K_u$	$0.85T_u$	–
PID	$0.6K_u$	$0.5T_u$	$0.12T_u$

good disturbance behavior is the design objective. For this purpose, the control loop should oscillate with a damping of $\vartheta \approx 0.2$ after about three periods. Per period the amplitude should decrease to $\frac{1}{4}$ of its previous value.

Method 1: Oscillation test In the oscillation test, the unknown system is supplemented with a P-controller the gain of which is increased step by step until the oscillation of the controlled variable changes to a continuous oscillation (see Fig. 6.64). A continuous oscillation is a critical and thus unstable (or marginally stable) state, but it enables the set controller gain K_u (also: critical gain) and period duration of the continuous oscillation T_u to be determined. From these values, a standardized conversion table can be used to calculate the final controller parameters for the various possible controllers, such as Table 6.3. This method has the advantage that no explicit system model is required, i.e. a potential controller can be attached directly to an unknown system. It can only be used for this purpose if the system itself is always stable, or if instability or operation in the marginal domain cannot lead to its damage/destruction. It is also quite possible that the controller found cannot meet the system specifications or that such a controller is not technically feasible.

Method 2: Step response measurement This method makes use of the fact that a large proportion of the systems to be controlled can usually be modeled sufficiently accurately as a P-T$_1$ element with additional delay or as a P-T$_2$ element. This is assumed to be a prerequisite for the system to be controlled. To find a suitable controller, the system gain, delay time and compensation time are determined from the step response of the system. The controller parameters can then be calculated again using Table 6.4, whereby the delay time T_L and the factor a from the step response diagram (see Fig. 6.65) of the system are used. Often you will also find the option to set the parameter a to 1 by default.

Here, too, it is possible that the controller found will not be technically feasible or will not satisfy the target conditions adequately. Depending on the application, other methods have also become established, such as the Chien/Hrones/Reswick method, but these will not be explained in detail here. Further information can be found, for example, in [1]

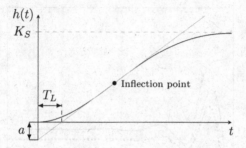

Fig. 6.65 Step response measurement for the parameter determination according to Ziegler–Nichols method 2

Table 6.4 Controller parameters for step response measurement

Controller	K_P	T_N	T_D
P	$1/a$	–	–
PI	$0.9/a$	$3.33T_L$	–
PID	$1.2/a$	$2T_L$	$0.5T_L$

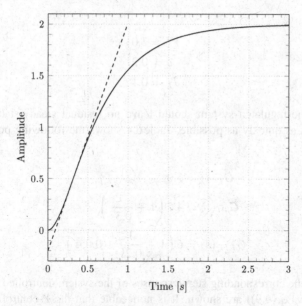

Fig. 6.66 Step response of the unknown example system

Here, the application of the Ziegler and Nichols method is presented as an example using the step response of an unknown system. The step response shown in Fig. 6.66 is first obtained.

From this step response, method 2 gives

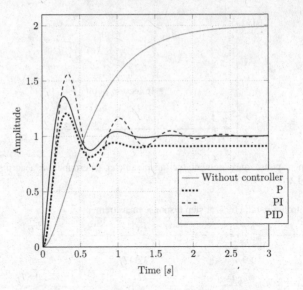

Fig. 6.67 Step response of the controlled example system in the feedback loop

$$K_s = 2.0$$
$$a = 0.2$$
$$T_L = 0.1$$

Ultimately, the regulated system should leave no residual steady-state error and should adjust as quickly as possible. Table 6.4 shows the following possible controllers.

$$G_P(s) = 5.0$$
$$G_{PI}(s) = 4.5 \left(1 + \frac{1}{0.3s}\right)$$
$$G_{PID}(s) = 6 \left(1 + \frac{1}{0.2s} + 0.05s\right)$$

In Fig. 6.67, the corresponding step responses of the system controlled in this way (according to Eq. (6.9)) are shown. It is noticeable that the P-controller leaves a permanent control difference, which means that it does not fulfil the requirement of stationary accuracy. The PI controller has stationary accuracy, but has a higher control rise and settle time than the PID controller and a higher overshoot range. The PID controller would therefore be the best choice for this system.

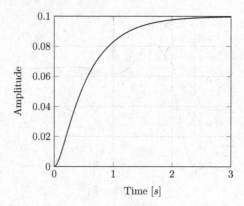

Fig. 6.68 Step response of a DC motor without controller

6.2.8.1 Speed Control of a Direct Current Motor According to the Ziegler–Nichols Method

Another example is the control of a DC motor, as described in Sect. 6.2.7.2. The output speed must be controlled. The system should run stably in the end, have as few fluctuations as possible with as little amplitude as possible and react as quickly as possible to changes in the reference variable. Since the speed is proportional to the applied voltage, it serves as the input variable.

The step response procedure is to be used again. Now there are two options. As in the example in Sect. 6.2.7.2, a mathematical model of the motor and the load can first be set up and used to calculate the transfer function and step response of the system. Alternatively, it is also possible to generate a step of the input variable of the real system and measure the "real" step response. The step response for this system can be seen in Fig. 6.68.

In order to make meaningful statements about the system properties, the target parameters must be converted into concrete numerical values. In our case this means that the settling time should be $T_{Off} < 2$ s, the overshoot should keep below $X_{OS} < 5\%$ and the permanent control difference should be $e(\infty) < 1\%$ of the amplitude.

The step response gives the values of the system without controller:

- The settling time: 3 s
- The final output value: 0.1
- The overshoot: 0%

This means that the controller must reduce the control difference and accelerate the system noticeably, but must not generate large overshoots.

Since no stationary accuracy is required, a P-controller could be sufficient, which should significantly shorten the settling time while causing no further overshoot. If this does not meet the requirements, a PID-controller is recommended, which also offers high speed and stationary accuracy, but allows overshooting. A pure PI-

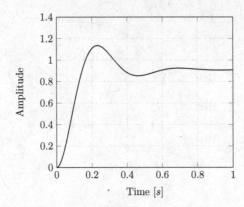

Fig. 6.69 Step response of a DC motor with a P-controller

controller does not seem advisable due to the expected delay resulting from the I element.

A P controller is tried first. Since the maximum exponent in the denominator of the transfer function of the system is equal to 2, it can be regarded as a P-T$_2$ system and therefore the Ziegler–Nichols method (method 2) can be used for parameter selection. Using Table 6.4, this results in a gain of $K_P = 100$. It can be seen from the step response of the system equipped with this controller (Fig. 6.69) that there is still a stationary control difference which is greater than the required value of <1%. The settling time is very good at <1 s. However, the overshoot range of approx. 18% is clearly too high. Since a further increase in gain would reduce the control difference but increase the overshoot, this controller is not suitable.

A PID controller is therefore considered next. According to the Ziegler–Nichols method, for the PID controller, $K_P = 120$, $K_I = 600$ and $K_D = 6$ result. As can be seen from the step response (Fig. 6.70) of the system created in this way, this system has an overshoot range of more than 5%, which is due to the large value of K_I.

If this value is reduced experimentally to 200, the step response shown in Fig. 6.71 is obtained. The overshoot range is clearly smaller, in addition the settling time was shortened further. However, the overshoot is still just above the required 5%.

Instead of a further reduction of K_I, K_D can also be increased. This also counteracts the overshoot, but accelerates the settling time. Increasing K_D to 10 gives the following values for the PID-controller

$$K_P = 120$$
$$K_I = 200$$
$$K_D = 10$$

As can be seen from the resulting step response (Fig. 6.72), this controller even exceeds all requirements and can be used for the system.

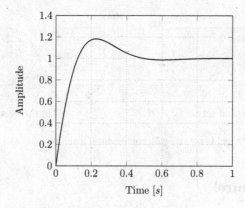

Fig. 6.70 Step response of a DC motor with a PID-controller

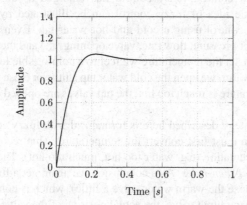

Fig. 6.71 Step response of a DC motor with a PID-controller, adapted integrative gain

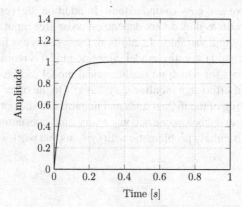

Fig. 6.72 Step response of a DC motor with a PID-controller, adapted integrative and differential gain

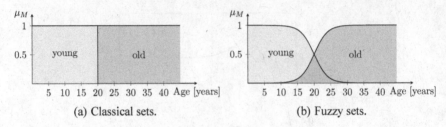

(a) Classical sets.   (b) Fuzzy sets.

Fig. 6.73 Classical sets and fuzzy sets with continuous transitions

6.3   Fuzzy Control

Alongside classical control, fuzzy control has emerged as an intuitive control approach. The basic idea of fuzzy control can be illustrated by the example of water temperature control using a cold and hot water tap. Even without a model of the system (water pressure, flow rate, valve opening, …) and the resulting control parameters, we can set the temperature we feel to a comfortable level, for example. If the water is too warm, we open the cold water tap a little or close the hot water tap. If the water temperature is much too hot, the tap valves are opened or closed slightly more.

The control strategy described here is formalized with fuzzy control. As shown in the example, we must first convert the temperature value recorded by a sensor into a fuzzy representation (e.g. water too hot, much too hot). The control strategy is described using *if-then-rules*. As a conclusion we again get a fuzzy value (e.g. *if* water is too hot, close the warm water valve a little), which is converted in the last step back into a crisp output value, the actual position of the valve.

Fuzzy control has several advantages. Control using fuzzy input values and if-then rules is very intuitive and easy to understand. In addition, the resulting controllers are usually very robust, as they do not depend on noise-free input variables and still deliver continuous output variables. In addition, the systems can be easily expanded without having to rebuild the entire model. Since no model is required, fuzzy control is suitable for systems for which no mathematical model can be derived or such a model requires great effort (e.g. nonlinear systems). On the other hand, in complex systems correct setting of the if-then rules and in particular the optimal value ranges of the fuzzy sets presented below can become a very time-consuming and error-prone task. This can be particularly problematic with systems that work with high recurring precision.

6.3.1   Basics of Fuzzy Set Theory

Fuzzy control is based on *fuzzy set* theory. In contrast to classical set theory, where an element either belongs to a set or does not (Fig. 6.73a), fuzzy sets define continuous

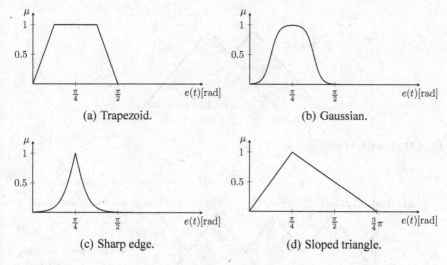

Fig. 6.74 Examples of membership functions

transitions (Fig. 6.73b). A person of the age of e.g. 20 belongs to both the *young* set and the *old* (to a certain extent). For this purpose, the characteristic function $\mu_M(x) \in [0, 1]$ of a fuzzy set M is introduced, which is also referred to as a membership function with continuous membership values.

A *fuzzy set* M with $M = \{(x, \mu_M(x)) : x \in X, \mu_M(x) \in [0, 1]\}$ assigns a degree of membership $\mu_M(x)$ to each element x of the basic set X. A normal, crisp set is therefore a special case of fuzzy sets. Frequently used membership functions are shown in Fig. 6.74. The degree of membership is not a measure of probability, as it is generally the case that

$$\sum_{x \in X} \mu_M(x) \neq 1 \qquad (6.70)$$

For *fuzzification*, i.e. assignment to the specific fuzzy sets, a crisp input value is generally given. This value is applied to its membership of the fuzzy sets. The example in Fig. 6.75 illustrates this. Two people aged 15 and 30 are taken:

$$\mu_{old}(15) = 0 \quad \mu_{young}(15) = \frac{2}{3}$$

$$\mu_{old}(30) = 1 \quad \mu_{young}(30) = \frac{1}{3}$$

Translated into fuzzy statements, this means that a 15-year-old is definitely not old, but not necessarily still young; a 30-year-old may still be considered young under certain circumstances. The parameter "age" is an example of a *linguistic variable*. They are used to reproduce expert knowledge in natural language. Each linguistic variable contains a set of *linguistic terms*, i.e. fuzzy values that the linguistic variables

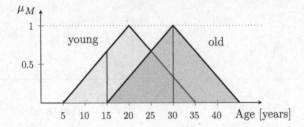

Fig. 6.75 Example of fuzzification

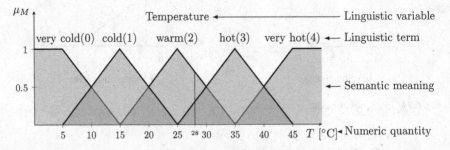

Fig. 6.76 Linguistic variables

can assume. These terms in turn represent fuzzy sets. In this example, the linguistic variable "age" contains the linguistic terms "young" and "old".

In the temperature control example, the linguistic variable "temperature" would contain the linguistic terms "very cold", "cold", "warm", "hot" and "very hot", as shown in Fig. 6.76. These are often converted into numerical values (here: 0–4) for reasons of presentation.

The semantic meaning is determined by the choice of the corresponding membership function. This is subjectively modeled by the expert. Linguistic operators (AND, OR, NOT) can be used to combine linguistic variables into linguistic expressions. In fuzzification, the crisp input values for each linguistic variable are transformed into the membership space of the linguistic terms involved. With n terms, the n-dimensional *sympathy vector* $s(x) = (\mu_1(x), \mu_2(x), \ldots, \mu_n(x))$ comes about. For each linguistic term A_i, the membership value $\mu_{A_i}(e)$ of the crisp input value e is determined. These values are combined to form the sympathy vector. This can also be seen in Fig. 6.76. For a temperature of 28°, the following applies.

$$\mu_{warm}(28°) = 0.7$$
$$\mu_{hot}(28°) = 0.3$$

This results in the sympathy vector

$$s(28°) = (0; 0; 0.7; 0.3; 0)$$

Table 6.5 Examples for t-norms and t-conorms

t-norm	Dual t-conorm
Minimum: $\min(\alpha, \beta)$	Maximum: $s(\alpha, \beta) = \max(\alpha, \beta)$
Lukasiewicz t-Norm: $\max(0, \alpha + \beta - 1)$	Lukasiewicz t-Conorm: $\min(1, \alpha + \beta)$
Algebraic product	Algebraic sum
$\alpha\beta$	$1 - (1 - \alpha)(1 - \beta) = \alpha + \beta - \alpha\beta$

Like fuzzy sets, there is also fuzzy logic. So you can arrive for an appointment at 5 o'clock only one minute late or be there ten minutes early. The greater the difference between the time of actual arrival and the agreed time, the less true (membership function) the statement that you are on time is.

Like Boolean logic for sets with binary membership functions, fuzzy logic determines axioms and operators for manipulating fuzzy sets. Fuzzy truth value functions should correspond with the classical binary truth value functions and be restricted to the extreme values 0 and 1. The following axioms apply:

(T1) $\quad t(\alpha, \beta) = t(\beta, \alpha)$ $\qquad\qquad$ Commutativity
(T2) $\quad t(t(\alpha, \beta), \gamma) = t(\alpha, t(\beta, \gamma))$ $\quad$ Associativity
(T3) $\quad \beta \le \gamma \Rightarrow t(\alpha, \beta) \le t(\alpha, \gamma)$ $\quad$ Monotony
(T4) $\quad t(\alpha, 1) = \alpha$

For all α, β, γ in $[0, 1]$. Complement, intersection and union in fuzzy sets correspond in Boolean logic to negation, conjunction and disjunction. Conjunction is defined using the *t-norm* (triangular norm). A function $t : [0, 1]^2 \to [0, 1]$ is called the t-norm if it fulfills the axioms (T1) to (T4). Examples of t-norms include:

$t(\alpha, \beta) = \min(\alpha, \beta)$ $\qquad\qquad$ Minimum
$t(\alpha, \beta) = \max(\alpha + \beta - 1.0)$ $\quad$ Lukasiewicz t-norm
$t(\alpha, \beta) = \alpha \cdot \beta$ $\qquad\qquad\quad$ Algebraic product

The conjunction is determined by $(\mu_1 \cap \mu_2)(x) = t(\mu_1(x), \mu_2(x))$.

The *s-norm* or *t-conorm* describes the disjunction. In analogy to the DeMorganian laws of Boolean logic, there is a dual connection between t-norms and t-conorms. Each t-norm t induces a t-conorm s by means of $s(\alpha, \beta) = 1 - t(1 - \alpha, 1 - \beta)$. Conversely, from a t-conorm s using $t(\alpha, \beta) = 1 - s(1 - \alpha, 1 - \beta)$ the corresponding t-norm t can be obtained. Examples of t-norms and the associated dual t-conorms can be found in Table 6.5.

To determine the union using the s-norm, it is used as follows.

$$(\mu_1 \cup \mu_2)(x) = s(\mu_1(x), \mu_2(x))$$

The complementary function can be calculated by $\mu^C(x) = 1 - \mu(x)$. The graphical solution is shown in Fig. 6.77.

A comparison between the classical and fuzzy truth value functions can be seen in Table 6.6.

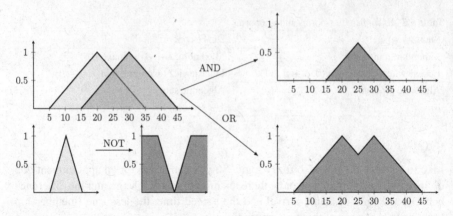

Fig. 6.77 Intersection, union and complement

Table 6.6 Classical and fuzzy truth functions

	Classical	Fuzzy
Intersection	$x \in M_1 \cap M_2 \Leftrightarrow x \in$ $M_1 \wedge x \in M_2$	$(\mu_1 \cap \mu_2)(x) =$ $t(\mu_1(x), \mu_2(x))$
Union	$x \in M_1 \cup M_2 \Leftrightarrow x \in$ $M_1 \vee x \in M_2$	$(\mu_1 \cup \mu_2)(x) =$ $s(\mu_1(x), \mu_2(x))$
Complement	$x \in \overline{M} \Leftrightarrow \neg(x \in M)$	$\mu^C(x) = 1 - \mu(x)$

The intersection and the union always work with the same basic sets. In order to combine fuzzy sets of different basic sets, the Cartesian product is introduced. Let $A_1, A_2, \ldots, A_n$ be fuzzy sets of the basic sets $X_1, X_2, \ldots, X_n$. Then the Cartesian product is (fuzzy relation)

$A_1 \times A_2 \times \cdots \times A_n$ from $X_1 \times X_2 \times \cdots \times X_n$ with

$\mu_{A_1 \times A_2 \times \cdots \times A_n}(x_1, x_2, \ldots, x_n) = t(\mu_{A_1}(x_1), \mu_{A_2}(x_2), \ldots, \mu_{A_n}(x_n))$

As shown in our introductory example, fuzzy control uses if-then rules. Based on a fuzzy state description (*premise*), fuzzy *conclusions* are generated by the if-then rules (*implication*). This transformation is also known as *fuzzy inference*. The implication is given as a fuzzy relation $\mu_R(x, y)$, $(x, y) \in X \times Y$ and the premise as a fuzzy set $\mu_P(x)$, $x \in X$. The conclusion is again a fuzzy set $\mu_{res}(y)$, $y \in Y$.

The conclusion is determined using the t-norm and the s-norm.

$$\mu_{res}(y) = s_{x \in X}\{t(\mu_P(x), \mu_R(x, y))\} \qquad (6.71)$$

If the frequently used minimum is applied for the t-norm and the maximum for the s-norm, the conclusion can be determined to

$$\mu_{res}(y) = \max_{x \in X}\{\min(\mu_P(x), \mu_R(x, y))\} \qquad (6.72)$$

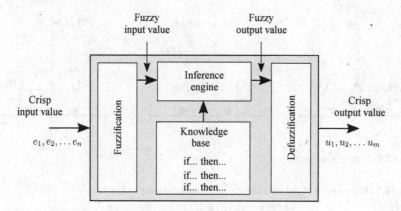

Fuzzy
input value

Fuzzy
output value

Crisp
input value

$e_1, e_2, \ldots e_n$

Inference
engine

Knowledge
base

if... then...
if... then...
if... then...

Crisp
output value

$u_1, u_2, \ldots u_m$

Fig. 6.78 Components of fuzzy control

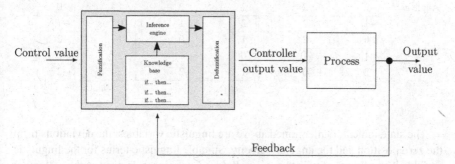

Control value

Inference
engine

Knowledge
base

if... then...
if... then...
if... then...

Controller
output value

Process

Output
value

Feedback

Fig. 6.79 Control process when using a fuzzy controller

6.3.2 Structure of a Fuzzy Controller

Formally, fuzzy control represents a static, non-linear mapping of input variables $e_i \in E_i$ onto the output variables $u_i \in U_i$. The input and output variables are exact/crisp values, as in classical control. The calculation of the controlled variable is divided into the following three steps (Fig. 6.78): 1. Fuzzification, 2. Fuzzy inference, 3. Defuzzification.

First, the crisp input value is converted into a fuzzy value. This fuzzy value serves as a premise for the inference. The blurred conclusion is determined using the if-then rules. The last step is *Defuzzification*, in which a crisp control value is calculated from the fuzzy conclusion. Fuzzy control differs from crisp control only in the replacement of the controller (and the calculation of the difference), as shown in Fig. 6.79.

Using the example of the inverse pendulum (Fig. 6.80, see also Sect. 6.2.7.3), the 3 steps are explained in more detail. The angular deviation from the zero position $e(t)$ [rad] and the angular velocity $\dot{e}(t)$ [rad/s] are selected to describe the state of the rod. The controlled variable is the force $u(t)$ [N] which acts on the carriage to bring the rod to the zero position or to hold it there.

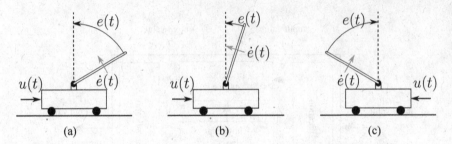

Fig. 6.80 Inverse pendulum

Table 6.7 If-then rules of the inverse pendulum corresponding to Fig. 6.81

Force		Angular velocity $\dot{e}$				
u		−2	−1	0	1	2
Angle e	−2	2	2	2	1	0
	−1	2	2	1	0	−1
	0	2	1	0	−1	−2
	1	1	0	−1	−2	−2
	2	0	−1	−2	−2	−2

The state indicators mentioned above are linguistic variables: the deviations from the zero position and the angular velocity. Suitable linguistic terms for the linguistic variable "angular velocity" are e.g. *small, large, very large*. Instead of the colloquial terms, standard definitions are usually used for linguistic terms such as negative-large (−2), negative-small (−1), ..., positive-large (+2) (see Fig. 6.81).

The actual control behavior is described in the if-then rules. Suitable rules for the system in Fig. 6.81 would be for example

- IF $e(t) = 0$ AND $\dot{e}(t) = 0$ THEN $u(t) = 0$
- IF $e(t) = 0$ AND $\dot{e}(t) = +1$ THEN $u(t) = -1$

The complete rules are shown in Table 6.7. The corresponding membership, as represented in Fig. 6.81 as thick bars, would be

$$e(t) = 0$$

and

$$\dot{e}(t) = \frac{\pi}{8} - \frac{\pi}{32} = \frac{3}{32}\pi \approx 0.294$$

In general, all rules that use linguistic terms in their IF rules whose memberships of the crisp input values are greater than zero are active.

The reliability with which a rule applies is defined by the relation of the linguistic terms involved. For a rule of the form IF $E_1 = A_1$ AND ...AND $E_n = A_n$ THEN

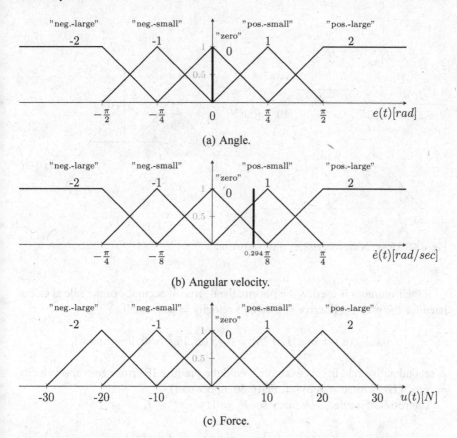

(a) Angle.

(b) Angular velocity.

(c) Force.

Fig. 6.81 Linguistic variables at the example of the inverse pendulum: **a** Angle, **b** Angular velocity and **c** Force

$U = B$, the degree to which the rule applies to a given input vector $e' = (e'_1, \ldots, e'_n)$ is

$$\mu_{A_1 \times \cdots \times A_n}(e'_1, \ldots, e'_n) = \mu_{\text{premise}}(e'_1, \ldots, e'_n)$$
$$= t(\mu_{A_1}(e'_1), \ldots, \mu_{A_n}(e'_n))$$

The following applies to the example of the inverse pendulum. If $e(t) = 0$ and $\dot{e}(t) = \frac{\pi}{8} - \frac{\pi}{32} \approx 0.294$ are the state description of the rod, then the membership functions specified above are used to give

$$\mu_{\text{zero}}(e(t)) = 1 \quad \text{and} \quad \mu_{\text{pos.-small}}(\dot{e}(t)) = 0.75$$

(Fig. 6.81).

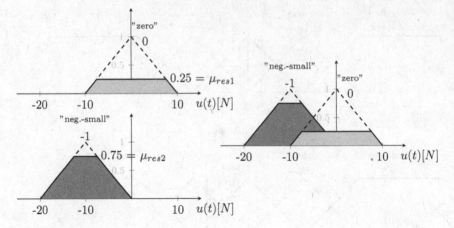

Fig. 6.82 Union of the conclusions of two active rules

If the minimum is used as the t-norm, the degree of accuracy of the rule is calculated for the premise IF error = zero and velocity = pos.-small as

$$\mu_{\text{premise}}(e') = \min\{1, 0.75\} = 0.75 \text{ (cf. Fig. 6.86, lower part.)}$$

A second active rule in this example would the premise IF error=zero and velocity = zero, THEN force = zero (cf. Fig. 6.86, upper part).

The result of a rule j is a fuzzy set B_j with

$$\mu_{res_j}(u) = t(\mu_{\text{premise}_j}(e'_1, \ldots, e'_n), \mu_B(u))$$

If you use the minimum as the t-norm, the result is

$$\mu_{res_j}(u) = \min\{\mu_{\text{premise}_j}(e'_1, \ldots, e'_n), \mu_B(u)\}$$

Figuratively speaking, the initial fuzzy quantity is capped at the level of the degree of fulfillment of the rule. In the next step, an OR operation (using the s-norm) for all conclusions of the active rules is performed, as shown in Fig. 6.82 for the example above.

$$\mu_{res}(u) = s(\mu_{res_1}(u), \ldots, \mu_{res_j}(u)) \overset{\text{e.g.}}{=} \max\{\mu_{res_1}(u), \ldots, \mu_{res_j}(u)\}$$

A crisp manipulated variable is generated from this resulting fuzzy quantity with the aid of defuzzification (Fig. 6.83). Different defuzzification methods are suitable for this purpose. For example, the maximum method returns the value with the largest membership.

$$u_0 := \max\{\mu_{res}(u) | u \in U\} \tag{6.73}$$

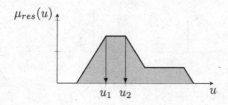

Fig. 6.83 Defuzzification

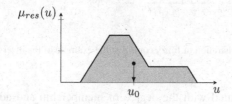

Fig. 6.84 Defuzzification: center-of-gravity method

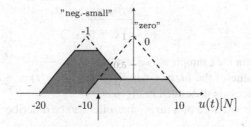

Fig. 6.85 Defuzzification: approximation

If there are several maxima $u_1, \ldots, u_m$, the maximum method can be extended as follows.

- **Left-Max-Method** $u_0 := \min\{u_1, \ldots, u_m\}$
- **Right-Max-Method** $u_0 := \max\{u_1, \ldots, u_m\}$
- **Mean-Max-Method** $u_0 := \sum_{i=1,\ldots,m} \frac{u_i}{m}$

Another defuzzification approach is the center of gravity method. The controlled variable belonging to the center of gravity of the resulting fuzzy set is returned (Fig. 6.84).

$$u_0 = \frac{\int_u u \cdot \mu_{ref}(u) du}{\int_u \mu_{ref}(u) du}$$

The disadvantage of this method is the high computational effort involved in the numerical integration. To avoid this, an approximation that is easy to calculate can be used (Fig. 6.85). Let u_i be the abscissa values of the centers of gravity of the conclusion sets. Since the membership functions are often triangular or trapezoidal functions, these values can be easily calculated. These abscissa values of the centers

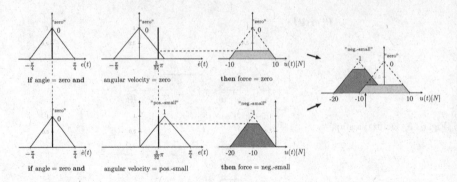

Fig. 6.86 Graphical illustration of fuzzy control at the example of the inverse pendulum

of gravity are weighted with the degree of membership and added together. The manipulated variable is therefore given by

$$u_0 = \frac{\sum u_i \mu_{\text{premise}_i}}{\sum \mu_{\text{premise}_i}}$$

Using this method for the example above results to $u_0 = \frac{-10 \cdot 0.75 + 0 \cdot 0.25}{0.75 + 0.25} = -7.5$. Thus the crisp output value of the fuzzy controller would be $u(t) = -7.5N$. The entire defuzzification process is illustrated in Fig. 6.86.

In summary, the structure of a fuzzy controller can be described in the following steps:

- Determine linguistic variables, terms and their membership function for the crisp input variables.
- Convert the crisp values into fuzzy values.
- Establish a set of IF THEN rules that describe the aspects of the control problem well.
- Select the IF THEN RULES that are active based on your premises and determine a set of fuzzy variables.
- Summarize these fuzzy variables and convert them into crisp output values.

References

1. Burns R (2001) Advanced control engineering. Butterworth-Heinemann, Oxford
2. Chen WK (ed) (2004) The electrical engineering handbook. Elsevier, Burlington
3. Michels K, Klawonn F, Kruse R, Nürnberger A (2007) Fuzzy control: fundamentals, stability and design of fuzzy controllers. Springer, Berlin

Part III
System Components of Embedded Systems

Chapter 7
Signal Processing

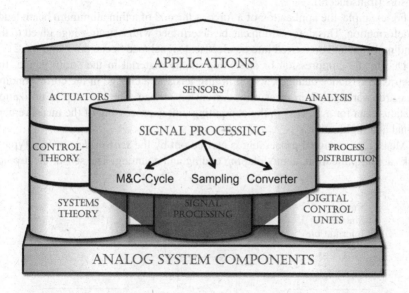

In this chapter the conversion between analog and digital signals (A/D or D/A conversion) is considered. The measurement and control cycle is introduced to link the work steps of the individual system components. Then the three phases of the A/D conversion with anti-aliasing filter, sample and hold circuit and the actual converter are described. The design of the associated parameters is based on the sampling theorem, which also plays a role in the design of communication channels (see Sect. 11.4.1). Then the D/A conversion with the corresponding circuits is introduced, also in three steps. Thus analog (sensor) signals can be converted into digital calculation values and digital control values can be turned back into analog values, e.g., to address actuators.

© The Author(s), under exclusive license to Springer Nature Switzerland AG 2021   197
K. Berns et al., *Technical Foundations of Embedded Systems*, Lecture Notes in Electrical
Engineering 732, https://doi.org/10.1007/978-3-030-65157-2_7

7.1   The Measurement and Control Cycle

Today's embedded systems are sometimes very complex and consist of a multitude of components. The previous chapters having dealt mainly with the electrical basics and principles of systems theory, the concrete "physical" structure of embedded systems and the processes running in them will now be discussed. In doing so, what has been learned so far is used to calculate and design the components, but also to correctly interpret and generate their signals. To this end, the process of signal processing will be examined in more detail below. With the help of the measurement and control cycle (see Fig. 7.1), the structure of an embedded system can be represented clearly.

In order to be able to influence any process, state variables must first be recorded. Examples of state variables are forces, velocities and positions. These variables are first converted into other physical variables with the aid of *non-electrical converters*, which can then be transformed into electrical variables using suitable measurement sensors (transducers).

For example, the application of a force at the end of a thin aluminum beam leads to deformation. This deformation can be determined with a strain gauge glued to the bending beam and converted into an electrical signal (see Sects. 8.6.1 and 8.6.2).

Due to the compression or stretching of the material in the strain gauge, the electrical resistance changes, which in turn leads to a change in the corresponding measured variable when measuring current or voltage. The last step in utilization of status data for a process in the computing unit is processing of the measurement signal and digitization.

Measurement signal processing is carried out by the *sensor electronics*. Typical tasks are amplification, demodulation, coding and counting. The sensor electronics

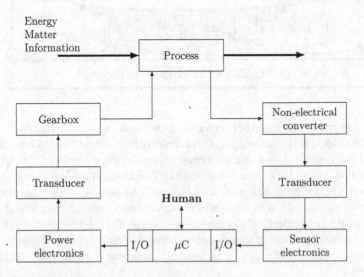

Fig. 7.1 Closed measurement and control cycle

can either output the status value as a digital value or prepare it for final processing with the help of an embedded computer. This conversion is called *A/D conversion* (also called *ADC, Analog/Digital Conversion*). The embedded computer, which can be a microcontroller or a digital signal processor (DSP), for example (see Chap. 10), then processes the digital data by algorithms such as controllers or filters.

An operator can influence the measurement and control cycle via a human-machine interface, for example by entering new system parameters. Since the actuators which are designed to change the process manipulate analog electrical voltages or currents, 'the control variables determined in the digital computer must first be converted. This conversion is called *D/A conversion* (or *DAC*). In order to map the analog control signals onto the working range of the actuators, these are adapted by means of the power electronics. These electrical quantities are converted into mechanical quantities with the aid of a conversion process and then adapted to the ambient conditions by means of gears or similar devices to enlarge a displacement.

If a DC motor is considered, it is controlled by current or voltage and generates a certain torque or angular velocity at the output of the shaft (see Sect. 9.3.1). The actuator can thus move an articulated arm. This disturbs the speed process into which the actuator is integrated. The action generated by the actuator or its effect is detected/measured again with sensors.

The elements mentioned create the "embedded system" only in conjunction. They are explained in more detail below.

7.2 Signal Processing

As shown above, much of the measurement and control cycle consists of recording and processing signals. For signal processing, analog signals must first be converted into digital values, then processed in the embedded computer and finally converted into (mostly analog) output variables with the help of a feedback process. This is done in several steps. First, the sensory input signal is converted into a digital signal by sampling. The purpose of sampling is to digitize the essential information content of the input signal completely. It consists of three consecutive steps (see Fig. 7.2 on the left):

- An anti-aliasing filter, implemented with an analog low-pass filter.
- A time quantization, performed with the help of a sample and hold circuit.
- An amplitude quantization for generation of a digital variable from the analog value.

The task of the return process is to convert a digital value into an analog output signal. Three phases are also involved here: The actual feedback, i.e. the conversion of the digital signal into an analog value, a sample and hold circuit used to reduce D/A conversion glitches and to improve the amplitude frequency response in data reconstruction. The output of the sample and hold circuit produces a staircase voltage which is smoothed out by a filter in the last step. Under the assumption that the system

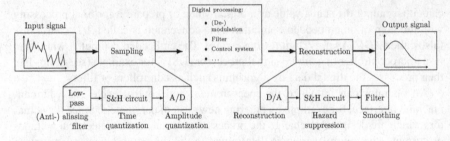

Fig. 7.2 Principle structure for digital processing of analog signals

algorithm operating in the embedded computer does not change the input signal and returns it directly, the output signal in an ideal system would be equal to the input signal. This is not the case in practice due to sampling, quantization and overdrive errors.

Resulting from these errors, the *signal-to-noise-ratio* (*SNR*) is used as a measure for the quality of the signal path (see also Sect. 11.4.1) as well as the single components. It is defined as the ratio of the signal power to the power of the noise, that is introduced by the environment and/or the components themselves. The SNR is usually given in decibel.

$$SNR_{dB} = 10 log_{10} \left(\frac{P_{signal}}{P_{noise}} \right) = 20 log_{10} \left(\frac{U_{eff,signal}}{U_{eff,noise}} \right)$$

The steps mentioned above will be examined in more detail below. Further information can be found in [1, 2] and [3].

7.3 Sampling

As shown above, in order to process an analog signal digitally, it must first be discretized in time and value, as shown in Fig. 7.3. This causes the continuous signal $s(t)$ to be converted into the discrete-time and discrete-valued sequence $\{s(nT)\}$ with $n = 0, \pm 1, \pm 2, \ldots$

Simplifed: $T = 1 \longrightarrow \{s(n)\}$ where $T = 1/f$ (sampling period)

For practical reasons, it is usually assumed that the sequences are finite.

$$\{s(n)\}_a^b = \{s(a), \ldots, s(b)\}, \text{i.e. } s(n) = 0 \text{ for } \begin{cases} n < a \\ n > b \end{cases} \tag{7.1}$$

The actual sampling is shown in Fig. 7.4. The sampled signal $s_3(t)$ is generated from the continuous-time signal $s_1(t)$ in the sampling unit by means of the sampled signal $s_2(t)$. Formally, this sampling unit represents a convolution of $s_1(t)$ and $s_2(t)$.

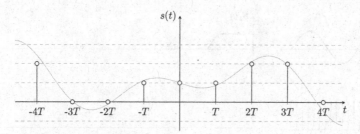

Fig. 7.3 Continuous and discrete signal

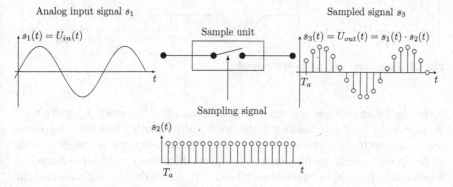

Fig. 7.4 Graphical illustration of sampling

7.3.1 Sampling Theorem

With the recorded analog signal, we are usually only interested in certain frequency components. Since, for example, the human ear can only perceive frequencies up to 20 kHz, it would not be advantageous for a sound recording to process significantly higher frequencies. On the other hand, real sensors are not able to pick up arbitrarily high signal frequencies, as explained below. It is therefore advisable to consider band-limited signals, i.e. signals whose frequencies lie within a fixed range. In Fig. 7.5 the functions in the time domain are compared with their Fourier transforms. Since the input signal is band-limited according to the specification, only spectral components up to the cut-off frequency f_{co} exist in the function $S_1(f)$. The spectrum of the sampling signal with the period $T_{spl} = 1/f_{spl}$ contains only frequencies that represent a multiple of f_{spl}. The spectral components $S_3(f)$ of the sampled signal can be determined by convolution of $S_1(f)$ and $S_2(f)$. The frequencies which are multiples of f_{spl} do not overlap. The continuous, band-limited signal $s(t)$ is transformed into an infinitely wide spectrum, periodic in f_{spl} in the frequency domain, by sampling. The side bands decrease with real rectangular sampling pulses.

In general, the question is now how to select f_{spl} in relation to f_{co} so that the continuous band-limited original signal can be reconstructed unambiguously.

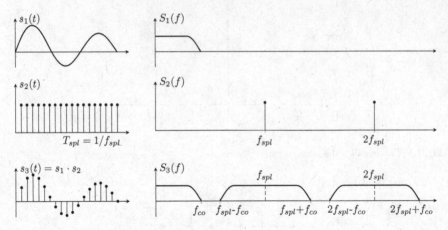

Fig. 7.5 Time course and spectra of the signals

In Fig. 7.6 an example is given in which the sampling frequency f_{spl} and cut-off frequency f_{co} were selected so that the periodically recurring spectral components are close to each other but do not overlap. If an overlap occurs, it is no longer possible to determine whether the frequency components still belong to the lower frequency band or already to the higher frequency band. This means that the original signal can no longer be reconstructed unambiguously. The Nyquist Shannon sampling theorem is derived from this consideration.

The sampling theorem states that a time function $s(t)$ with a spectrum in the interval $[0, f_{co}]$ is completely described by its sampled signal if the sampling frequency f_{spl} is more than twice as large as the cut-off frequency f_{co}. Mathematically speaking

$$f_{spl} > 2 \cdot f_{co} \tag{7.2}$$

The larger f_{spl} is in relation to f_{co}, the larger the spacings in the amplitude spectra. Since a high sampling frequency is technically more complex than a low one, it is in the developer's interest to limit the signal to the necessary frequencies. Low-pass filters are used for this purpose. At high sampling frequencies, a simple low-pass filter is sufficient to obtain non-overlapping spectra. In order to keep the sampling frequency as low as possible in relation to the cut-off frequency, an ideal low-pass

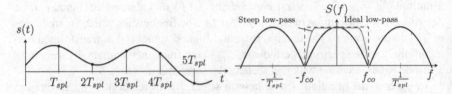

Fig. 7.6 Time domain and frequency spectra

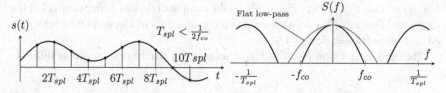

Fig. 7.7 Sufficiently fast sampling

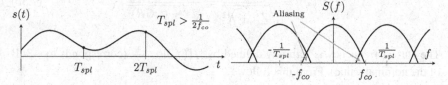

Fig. 7.8 Too slow sampling results in aliasing

filter whose edges drop off very steeply would be necessary (see Figs. 7.6 and 7.7). Note that this applies to both sides of the spectrum, i.e. to both positive and negative frequency components.

If the sampling frequency is too low, the output spectra overlap so that the original signal can no longer be reconstructed. This effect is called *aliasing* (see Fig. 7.8). Low-pass filters that try to prevent this are therefore called *anti-aliasing filters*.

7.3.2 (Anti-)Aliasing Filter

Anti-aliasing filters are analog low-pass filters that work either passively as RLC networks (see Sect. 3.5.1) or actively with additional operational amplifiers.

The following is an example of this. The RC low pass in Fig. 7.9 can be considered as a complex voltage divider. The following applies $U_{in}(t) = I_{in}(t) \cdot (R + \underline{Z}_C)$ and

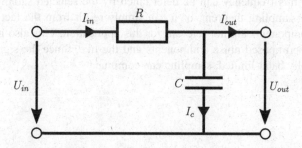

Fig. 7.9 Passive RC low pass filter

$U_{out}(t) = I_{in}(t) \cdot \underline{Z}_C$. Here $\underline{Z}_C = \frac{1}{j\omega C}$ is the complex resistance (*Impedance*) of the capacitor. This auxiliary value is used in the calculation of electrical networks with AC components (see Sect. 3.4.4).

The transfer function $H(\omega)$ and its absolute value can be determined as follows.

$$H(\omega) = \frac{U_{out}}{U_{in}} = \frac{\underline{Z}_C}{\underline{Z}_C + R} = \frac{1}{1 + j\omega CR} \tag{7.3a}$$

$$|H(\omega)| = \frac{|\underline{Z}_C|}{\sqrt{\underline{Z}_C^2 + R^2}} = \frac{1}{\sqrt{1 + (\omega CR)^2}} \tag{7.3b}$$

The cut-off frequency is formally defined by $|H(f_c)| = \frac{1}{\sqrt{2}}$ (corresponds to -3dB of the nominal value). From this follows

$$f_c = \frac{1}{2\pi RC} \tag{7.4}$$

With the help of this formula the low pass parameters R and C can be determined.

The following example illustrates the use and purpose of the upstream low-pass filter. The rectangular voltage signal $s(t)$ is to be sampled. Therefore

$$s(t) = \begin{cases} 5V & |t| \leq 1s \\ 0V & else \end{cases}$$

By Fourier transformation, in the frequency domain using Table C.2, we get

$$S(f) = 10Vs \cdot si(2\pi f \cdot 1s)$$

The problem is, that this function is defined for $-\infty \leq f \leq \infty$. So there is no natural maximum frequency. Each sampling will therefore result in various overlaps from which no clear signal can be fed back. The solution is to define a cut-off frequency beyond which the signal is cut off in the frequency domain by the low-pass filter, which serves here as an anti-aliasing filter.

In reality, this frequency can be determined by the selected sampling element based on the sampling theorem, or it can simply result from the fact that higher frequency components are not relevant for this application. This also includes, for example, superimposed noise components and the like. Since the signal has now been artificially band-limited, sampling can continue.

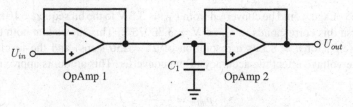

Fig. 7.10 Operational amplifiers to decouple the sample and hold circuit

7.4 Sample and Hold Circuit

Most A/D converters need some time to convert the sampled value into a digital signal. It is therefore necessary to keep the analog voltage constant during the conversion time. This is done using a sample and hold circuit as shown in Fig. 7.10.

The switch (which can take the form of a MOSFET, for example) is synchronized with the actual converter. Before conversion, it is closed so that the capacitor C_1 is charged to the applied voltage. During conversion, the switch is opened so that the subsequent signal change of U_{in} at the input does not influence the process.

The input and output sides are electrically decoupled by operational amplifiers, so that no charge can flow off here. The OpAmps are used as non-inverting amplifiers with an amplification of 1. As a result, they have a theoretically infinitely high input resistance. The left OpAmp is used to avoid influencing the input signal as far as possible. The second operational amplifier minimizes the discharge of the capacitor so that the applied voltage remains approximately constant over the conversion period.

However, this maintenance of the voltage value results in the problem that the sampling period must not be shorter than the holding period required for the conversion, as otherwise a new input value would overwrite the old value still being converted and thus falsify the result of the conversion. This time quantization is therefore determined in such a way that a maximum sampling is achieved, which is specified by the converter speed. This is the only way to ensure error-free sampling according to the sampling theorem. For the sampling rate, this gives

- **Max. Sampling rate** determined by converter speed,
- **Min. Sampling rate** determined by sampling theorem.

7.5 A/D Converters

The A/D converter converts the analog input signals into digital output signals, which are then processed by a digital computer. Depending on the requirements and the converter properties, various A/D converters are used in practice. In general, the applied voltage value is represented as a fraction of a reference voltage. If, for example, an A/D converter with a resolution of 8 bits had a reference voltage of $U_{Ref} = 5\,\text{V}$, the sampled voltage would be converted from $U_{in} = 3.3$ V to the bit sequence 10101001. Vice versa this corresponds to $\frac{169}{256} \cdot 5$ V $= 3.30078$ V. This shows that both the resolution (e.g., for $n = 8$ bits a resolution of $2^n = 256$ steps) and the level of the reference voltage affect the accuracy of the converter. This amounts approximately to

$$p_{min} = \frac{U_{Ref}}{2^n}$$

A distinction is made between direct and indirect converters. Unlike indirect converters, direct converters do not require an intermediate value. Examples of some of the most important converters are described below. Examples of direct converters include the parallel converter, the scaling method and the priority encoder. Examples of indirect converters are multi-slope, charge-balancing and delta sigma converters.

7.5.1 Successive Approximation Converter

The *Successive Approximation Converter* belongs to the class of direct converters. It compares the input variables step by step with certain reference variables. In the example shown, a 3-bit A/D conversion is performed. First, the input voltage is compared with half the reference voltage (maximum voltage value occurring). If the input voltage is higher, it is then compared with three quarters of the reference voltage. If this value is smaller, the interval is again halved, i.e. the input voltage is checked to determine if it corresponds to more than $\frac{5}{8}$ times the reference voltage. Figure 7.11 describes the procedure for further comparison results. At each comparison level, a new bit is determined, starting with the determination or calculation of the most significant bit (MSB). The structure of an A/D converter with scaling method can be seen in Fig. 7.12.

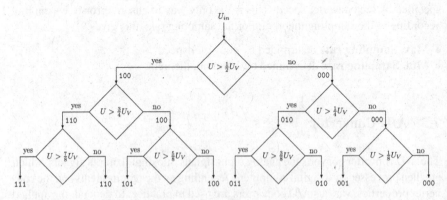

Fig. 7.11 Flowchart for the successive approximation conversion

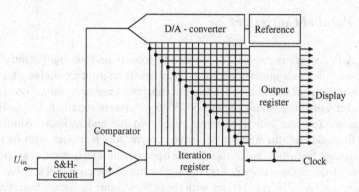

Fig. 7.12 Structure of a successive approximation converter

Here, the respective reference voltage is generated with the help of a D/A converter, which is compared by means of an operational amplifier as comparator. The comparison result in the iteration register determines the corresponding bit in the output register. This step-by-step determination of the digital word depends on the time for the desired resolution. The higher the resolution, the longer the conversion takes.

Successive approximation can only be used if the input voltage changes by a maximum of $\frac{1}{2}$ LSB during the conversion process. Therefore a sample and hold circuit is usually necessary. The process is also susceptible to aliasing, which is why a low-pass filter is often required. One clock period is required per bit, since a comparison is carried out per bit. Normally, a start and a stop clock cycle are required in addition. The conversion time T_c then results in

$$T_c = \frac{1}{f_{clock}}(n+2) \qquad (7.5)$$

with the bit width n.

The magnitude of the clock frequency depends mainly on the transient response of the comparator, which also affects the accuracy. It must have a high gain, since it usually has to supply a level of $5\,\text{V}$ at the output and the input voltage difference must be a minimum of $\frac{1}{2}$ LSB. The required gain is then

$$A = \frac{5\,\text{V}}{0.5 U_{LSB}} = \frac{5\,\text{V}}{0.5 U_{FS}} 2^n \qquad (7.6)$$

With a 16-bit converter and a measuring range of $2\,\text{V}$, the required gain is already more than 10^5. Only a few comparators offer these values and good transient response at the same time.

7.5.2 Parallel Converters

Parallel A/D converters are used to reduce the conversion time significantly. These are necessary, for example, for video applications. If an n-bit resolution of a digital word has to be generated, $2^n - 1$ analog comparators are necessary. For example, for a 16-bit word this would require 65535 comparators required. The individual reference voltages are generated via a voltage source and resistors. Similar to a clinical thermometer, the number of active comparators increases with increasing analog voltage from bottom to top. In the downstream electronics, a snapshot of the status of the comparators is first taken and stored for each clock pulse. In the subsequent encoder, the comparator with the highest value is given a binary code.

With parallel converters (also called *flash converters*), the voltage to be measured is compared simultaneously with several reference voltages. For this purpose, the valid measuring range is divided into 2^n intervals, whereby the accuracy is specified in n bits, and the input voltage is compared with each of these intervals. The $2^n - 1$ comparison results are converted into a binary value by a priority decoder.

The structure shown in Fig. 7.13a results in the measured value always being rounded down. According to the voltage divider rule (see formula 3.5), there is a voltage of $U_{Cmp_1} = \frac{1}{8}U_{Ref}$ at the leftmost resistor, of $U_{Cmp_2} = \frac{2}{8}U_{Ref}$ at the next resistor, and so on. An input voltage $U_{in} = 2.95V$ for a 3-bit parallel converter with a measuring range of 0 to 8V would thus be assigned the value 2 (corresponds to $2 \cdot \frac{8V}{2^3} = 2$ V) instead of the expected value 3. To avoid this, the resistors are adapted as in Fig. 7.13b. The voltage $U_{in} = 2.95V$ in the example would now be assigned the value 3, but voltages greater than 7.5 V would also be rounded off. In order to

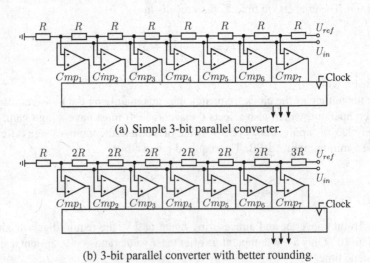

(a) Simple 3-bit parallel converter.

(b) 3-bit parallel converter with better rounding.

Fig. 7.13 Example circuits of parallel converters

recognize this case, a further comparator would have to be integrated which shows the overflow.

7.5.3 Priority Decoder

A priority decoder outputs the n-bit index of the most significant of the 2^n input bits, which has the value 1. In the case of parallel converters, the terms "thermometer decoder" or "thermometer ADC" are also used, since the height of the column of ones indicates the measured voltage. A simplified representation of usage can be seen in Fig. 7.14. Priority decoders can be used to evaluate a parallel conversion.

V_i is the result of the ith comparator. The AND gates ensure that only the input of the ROM that is located at the transition from the ones column to the zero column is active. If, for example, the voltage $U_{in} = 3.3V$ is converted, V_0, V_1 and V_2 have the value 1, all others have the value 0; thus only input 3 of the ROM is active. The number i is stored in binary form at address i; the ROM itself does not contain a decoder, the stored data words are addressed directly via the inputs.

This approach is also well suited to output the measurement result in a different form, e.g., in Gray coding or as BCD numbers. For this purpose, the values are stored in corresponding coding by the ROM; a conversion from the dual coding is therefore not necessary.

The transfer speed here is very high, as all comparisons are carried out simultaneously. This enables sampling frequencies of up to 2.2 GHz at 8-bit precision (e.g. with. Maxim MAX109). Such A/D converters are therefore used, for example, in radar receivers and in video technology. If even higher frequencies are required, the input signal must be distributed over several converters by means of a demultiplexer, which process the signal alternately.

However, parallel converters have two major disadvantages:

- Their size, i.e. the number of components, grows exponentially with accuracy. In addition, the large number of comparators leads to a high input capacitance, so that a high-quality amplifier must be connected upstream.
- The resistors for grading the reference voltage must be manufactured very precisely, both relative to each other and relative to the total resistance.

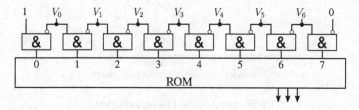

Fig. 7.14 Priority decoder

For these reasons, parallel converters are usually only offered up to 8 bits precision, rarely up to 10 bits. Higher accuracies cannot be produced commercially.

7.5.4 Dual Slope Method

Another class of A/D converters are the integrating converters. In contrast to the direct methods mentioned above, integrating A/D converters gradually increase or decrease voltages in order to approximate the input voltage. The number of time steps determines the digital value. An example of this is the dual-slope method, for which a circuit is shown in Fig. 7.15.

The dual slope method is divided into three phases. In the first phase, the converter is reset, i.e. the capacitor is emptied and the digital counter is reset to zero (the capacitor C_1 is discharged when S_2 is closed). The measurement itself, also known as the measurement cycle, takes place in the remaining phases.

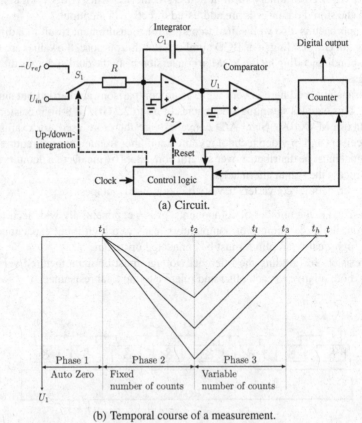

(a) Circuit.

(b) Temporal course of a measurement.

Fig. 7.15 Example of the dual slope method

In the second phase, the input voltage to be converted U_{in} is integrated with an operational amplifier via a fixed time $T_2 = t_2 - t_1$ (N clock pulses), so that the voltage $U_1(t_2)$ behind the integrator is proportional to the input voltage U_{in} (see Sect. 4.10.3).

$$U_1(t_2) = -\frac{1}{RC} \int_{t_1}^{t_2} U_{in} dt = -U_{in} \frac{t_2 - t_1}{RC} \tag{7.7}$$

When this downward integration is complete, a reference voltage $-U_{ref}$, which has the opposite sign of U_{in}, is gradually integrated up until the comparator determines 0V or a voltage opposite to the input voltage. The number n of clock pulses required for this is proportional to the input voltage, i.e. the output at the counter currently represents the digital word. Formally, the process can be described as follows.

Upward integration is carried out repeatedly with the reference voltage $-U_{ref}$ until $U_1 = U_1(t_3) \geq 0V$ is reached. Since this takes different lengths of time depending on $U_1(t_2)$, the number n of clock pulses required is variable.

$$U_1(t_3) = U_1(t_2) - \frac{1}{RC} \int_{t_2}^{t_3} -U_{ref} dt \tag{7.8}$$

For $U_1(t_3) = 0$ and $T_3 = t_3 - t_2$ the relationship of the voltages is

$$-U_{in} \frac{T_2}{RC} - (-U_{ref}) \cdot \frac{T_3}{RC} = 0 \tag{7.9}$$

Resulting in

$$U_{in} = -\frac{-U_{ref} T_3}{T_2} = -\frac{-U_{ref} nT}{NT} = -\frac{-U_{ref}}{N} n \tag{7.10}$$

(T=Duration of the period of the nominator)

Thus, the number of clock cycles n, which the upward integration takes is proportional to the input voltage.

$$U_{in} \sim n \tag{7.11}$$

In Fig. 7.15b, the course of the voltage U_1 during two example measurements is represented. In example 1 a higher voltage is converted, therefore the curve in the first section (from t_1 to t_2) is steeper than in example 2. The curves in the second section (t_2 to t_h in example 1 and t_l in example 2) are parallel to each other, since the reference voltage that is used for the upward integration is the same in both cases. A special feature of dual-slope converters is that interference frequencies (e.g., 50 Hz/60 Hz through the mains) are completely filtered out if such a frequency is a multiple of the inverse of the integration time T_2. For example, if the integration time of a transducer is $T_2 = 100\,ms$, in theory any frequency that is a multiple of

$\frac{1}{T_2} = 10\,\text{Hz}$ is filtered out. In practice, there are limits to this filtering, partly because interference frequencies are not perfect and are subject to fluctuations.

Dual-slope converters have a relatively high accuracy and are relatively robust, because errors due to manufacturing tolerances or temperature drifts are partially offset by the up and down integration. Nevertheless, precise components are required, especially in the integrator. Due to the robustness, only relatively little coordination between the components is required, which is also independent of the resolution. This enables cost-effective production.

The disadvantages are the relatively long conversion time and the limited scope of the resolution, because the duration of the measurement process increases exponentially with accuracy and is also dependent on the input voltage. For example, the TC510 from Mikrochip Technology Inc. achieves a sampling rate of only 4–10 values per second. The maximum architectural conversion frequency is a few kilohertz (e.g., 3 kHz at 14 bits: Texas Instruments ICL7135).

Commercial converters with the dual-slope method are available up to an accuracy of 17 bits (e.g., Microchip Technology Inc. TC510).

7.5.5 Multi-slope Converter

Since the end of the upslope integration phase is synchronized with a clock, dual-slope converters usually integrate for too long, so that a positive voltage U_1 is then applied behind the integrator. In order to increase the precision of the converter, this residual voltage can be amplified and integrated downwards.

Figure 7.16 shows an example of the course of a voltage U_1 during a measurement. The amplification of the residual voltage after the first conversion is not shown for reasons of clarity.

A voltage is converted with a precision of 10 bits, resulting in the 10 high-quality bits. The rest is amplified by the factor 2^5 and integrated downwards. Since there is

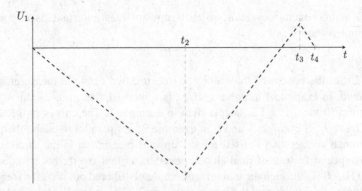

Fig. 7.16 Course of a multi-slope conversion

a remainder here again, it can be amplified by the factor 2^5 and integrated upwards again. The result of this measurement is calculated as follows.

$$(Result_1 \cdot 2^{10}) - (Result_2 \cdot 2^5) + (Result_3 \cdot 2^0) \tag{7.12}$$

This results in a precision of 20 bits, provided that the quality of the amplifier is high enough not to falsify the result. A further advantage is that a measurement requires only $2^{10} + 2^5 + 2^5 \approx 1,100$ steps instead of $2^{20} \approx 1,000,000$, i.e. the process is almost accelerated by a factor of 1000.

7.5.6 Charge Balancing Method

The charge balancing method (C/B method) is another integrating AD-converter. It belongs to the types of voltage-frequency converters (U/F converters). With these methods, the analog voltage is transformed into a pulse sequence with a constant voltage. This intermediate signal can be read directly into the evaluation electronics via a digital input.

The analog voltage value can be encoded in different ways. With the C/B method, the information is encoded in the frequency of the pulse sequence, as can be seen in Fig. 7.17. An integrator is charged with the sum of the input voltage U_{in} (*Charge*) and a reference voltage $-U_{Ref}$ for a fixed period of time T_s. This period of time is created e.g., by a monostable multivibrator (see Fig. 7.17). This is a circuit that, when triggered by a pulse or step at the input, creates a pulse of defined duration T_s at its output. After the end of this period of time, $-U_{Ref}$ is switched off the circuit, so that

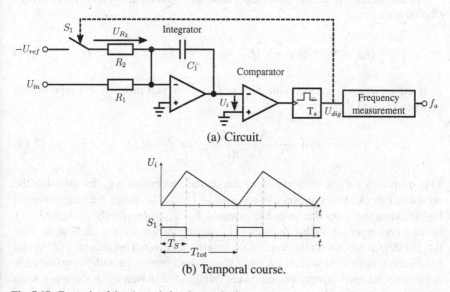

(a) Circuit.

(b) Temporal course.

Fig. 7.17 Example of the charge balancing method

U_{in} is integrated downward until a threshold value at the comparator is exceeded. This triggers a pulse at the output of the comparator that starts a new pulse in the monostable multivibrator and thus restarts the whole process. The higher the voltage to be measured, the faster the comparator threshold is reached again (high voltage $\Rightarrow$ high frequency).

The charges Q_{tot} on the capacitor are paramount for the voltage at the output of the integrator.

$$Q_{tot} = U \cdot C = Q_{ref} + Q_{in} = - \int\limits_{t_0}^{t_0+T_{tot}} \left(\frac{U_{in}}{R_1} + \frac{U_{R_2}}{R_2} \right) dt \qquad (7.13)$$

$$\text{with} \quad U_{R_2} = \begin{cases} -U_{ref}, & t_0 \leq t \leq T_S \\ 0, & else \end{cases}$$

Since the output frequency $f = 1/T_{tot}$ of the circuit is much higher than the signal frequency (oversampling), the voltage can be assumed to be constant for one cycle. Thus, the charge introduced to the capacitor by the input voltage U_{in} over one cycle is

$$Q_{in} = -\frac{U_{in}}{R_1} T_{tot}$$

The charge introduced by the reference voltage $-U_{ref}$ is

$$Q_{ref} = \frac{U_{ref}}{R_2} T_S$$

The aim of the circuit is to balance the charges on the capacitor on average to zero (*Balancing*).

$$Q_{tot} = 0 \Rightarrow Q_{in} = -Q_{ref} \Rightarrow -\frac{U_{in}}{R_1} \cdot T_{tot} = -\frac{U_{ref}}{R_2} \cdot T_S \qquad (7.14)$$

If the equation is solved with $f = 1/T_{tot}$, there is a proportional relationship between U_{in} and f.

$$f = \frac{R_2}{U_{ref} \cdot T_S \cdot R_1} \cdot U_{in} \Rightarrow f \sim U_{in} \qquad (7.15)$$

This frequency can be determined by the digital electronics e.g., by counting the number of clock-cycles between two pulses of U_{dig}. The charge balancing method has the advantage over the multi-loop method that U_{in} is constantly connected over the conversion period, so that changes of U_{in} during the conversion can be accounted for. The decisive factor for the result of the measurement is not the accuracy of a cycle, but the average time of several cycles. An advantage over the parallel approaches is the simpler design with only one reference voltage. A disadvantage is the much more complex digital circuit electronics, since the individual cycles run with a multiple of the signal frequency. The circuit in Fig. 7.17 shows an asynchronous C/B circuit, as it is not synchronized with a clock. By exchanging the monostable multivibrator with a clocked Flipflop (see Sect. 7.5.7), it could be synchronized.

7.5.7 Delta Modulation

With the delta converter, the converter output signal encodes a difference in the input signal between two sampling points. The steeper the edge of the input signal, the more pulses are output with the corresponding sign. With DC voltage, there is a constant alternation of logical ones and zeros. The basic idea is that the signal to be converted usually has high data redundancy. This means that neighboring samples differ only slightly from each other. This has the advantage that data reduction is possible, which is important for higher sampling rates (e.g., for audio and video). The principle is shown in Fig. 7.18. From an input signal, the integral of the previous pulses of U_{dig}, which act as estimation, is subtracted. This difference is then quantized to get a digital signal. A possible realization is shown in Fig. 7.19. The block denoted as "FF" is a *Flipflop*. This is a digital circuit, that—in this case—delays the (binary) input value (here: U_V) for one cycle of the clock input (here: $Clock$). Additionally, it only changes it's output value at the rising edge of the clock signal. Thus, its output signal is digital (discrete-time and discrete-valued). It has to be noted, that the "logical zero" at the output of the flipflop has to be electrically negative to assure a correct integration.

Redundant signal components (= signal components that can be reconstructed over the probable signal curve) are determined by a predictor and not quantified. Only the remaining signal component (difference between current value and predicted value) must be coded. In the simplest case, the predicted value is the signal value determined by the last sample. The difference results in a 1-bit coding of the

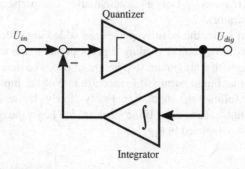

Fig. 7.18 Principle structure of the delta modulation

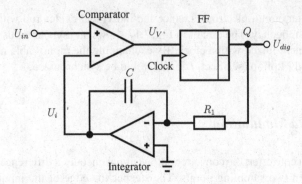

Fig. 7.19 Example circuit for the delta modulation method

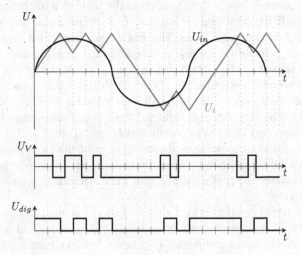

Fig. 7.20 Temporal course of the delta modulation of a sine-like signal

difference between U_{in} and U_i. Only a "larger/smaller" comparison of the two values is necessary (comparator).

In the evaluation logic, the positive pulses are added together and the negative values are subtracted. The counter reading is proportional to the input voltage. This results in a staircase with a maximum step height of one. The maximum step height of one results in a maximum permissible increase rate of the input signal. A small analog step height follows the input signal only slowly. If the step is too small, gradient overload (noise) can occur. If the steps are too large, the quantization noise increases. This is also depicted in Fig. 7.20.

7.5.8 Delta Sigma Converter

The delta sigma converter is based on the principle of delta modulation. While the delta modulation's output encodes the difference of the input value from the esti-

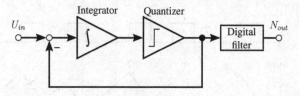

Fig. 7.21 Principle structure of a delta sigma converter

mation, the delta sigma converter encodes an equivalent of the input itself. This is in principle achieved by adding an integration or summation stage, as shown in Fig. 7.21. If a constant input voltage is present, a sequence of pulses results at the output. The average of these pulses over a given period of time equals the input voltage. If the voltage increases, the number of pulses during the same period of time increases, too. If the voltage decreases, the number of pulses during this period of time decreases, too. This is shown in Fig. 7.23.

In general, the circuit of a delta sigma AD-converter is similar to the charge balancing converter, as depicted in Fig. 7.22. The main differences are the Flipflop, that is used to synchronize the circuit to a fixed clock and the digital filter, that "interprets" the output values. As the output signal U_{dig} is already a series of pulses with the time between the pulses being proportional to the height of the input voltage, it can also be interpreted differently. By counting the number of pulses over a given period of time instead of determining the time between those pulses, a constant update frequency can be obtained. Additionally, the relative measurement error (cf. Sect. 8.2) becomes independent of the input signal this way.

As an example, a delta-sigma converter with a reference voltage of 2.5 V is assumed and the input voltage is $U_{in} = 1$V, then the values shown in the Table 7.1 and Fig. 7.23 result for the first steps. The result is the sequence 010010100101001..., so two of five values are ones, i.e. the input value corresponds to $\frac{2}{5}$, resp. 40% of the full input range. The input range is from $0V$ to U_{ref}, thus the decoded value would be 1 V. The precision of the result now correspods to the width of the period of time used for evaluation. As shown in Fig. 7.23, a period of 5 cycles results in the correct result after it is settled, while a period of six cycles results in the output oscillating

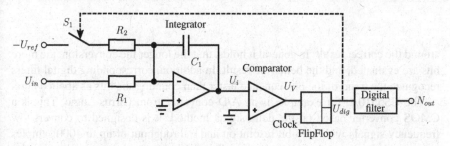

Fig. 7.22 Example circuit for a delta sigma modulator

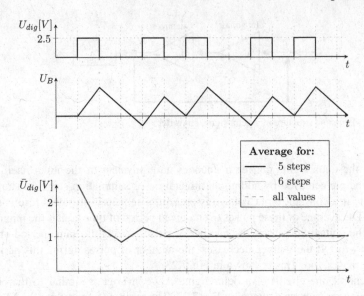

Fig. 7.23 Temporal course of the digital output, the integrator output and the average of the digital output for different averaging periods

Table 7.1 Delta sigma conversion with $U_{Ref} = 2.5V$ and an input voltage of $1V$

Step	Switch	Comparator	U_{dig}
1	(open)	1	2.5
2	closed	0	0
3	open	0	0
4	open	1	2.5
5	closed	0	0
6	open	1	2.5
7	closed	0	0
8	open	0	0
9	open	1	2.5
10	closed	0	0
11	open	1	2.5

around the correct result. In general it holds, that the longer the conversion, the more bits are evaluated, and the better the result. In addition, corresponding digital filters recognize the trend of the bit sequence faster than counting the bits as applied here.

The CS5510A, for example, is an A/D converter from Cirrus Logic. This is a CMOS converter based on the delta sigma method. It is designed to convert low frequency signals with a 16-bit resolution and a throughput of up to 4000 samples per second. The clock is supplied externally. The integrated low-pass filter enables cut-off frequencies between 0.1 and 10 Hz.

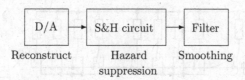

Fig. 7.24 Reversal: D/A converter, sample and hold circuit and filter

7.6 Reversal Process

In order to address the actuators in an embedded system, it is generally necessary to convert the calculated digital output quantities into analog electrical quantities. This process is called the reversal process.

This process is divided into three consecutive phases: the actual D/A conversion, sampling of the analog value at the output of the D/A converter and holding of the converted value over a clock period using a sample and hold circuit and smoothing of the output signal by means of a low-pass filter (see Fig. 7.24).

7.7 D/A Conversion

7.7.1 Parallel Converters

First, the actual D/A conversion is considered. In principle, the process of A/D conversion is reversed here. An analog signal is generated from a digital bit sequence. The methods used are therefore often comparable. Parallel D/A converters carry out a synchronous conversion of the digital code into analog values. A possible circuit is shown in Fig. 7.25.

The individual bits in the output register switch the reference voltage U_{ref} to the weighting resistors at the input of an adder. The ratios of the partial currents/voltages

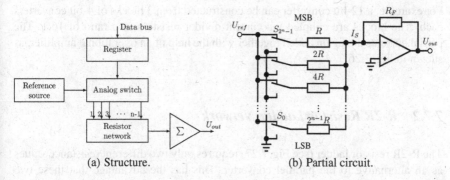

(a) Structure.   (b) Partial circuit.

Fig. 7.25 Example circuit for a parallel converter

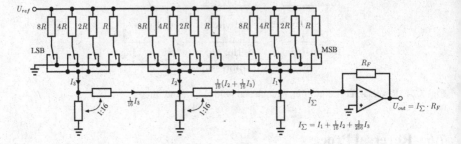

Fig. 7.26 Example circuit for a 12-bit parallel converter

correspond to the binary representation. The partial currents/voltages are added together at the input of the operational amplifier S and, if necessary, amplified for the application (adder circuit). The output U_{out} is proportional to the binary word $(b_{n-1}; b_{n-2}; \ldots; b_1; b_0)$ at the input.

$$I_S = b_{n-1} \cdot \frac{U_{ref}}{R} + b_{n-2} \cdot \frac{U_{ref}}{2R} + \ldots + b_0 \cdot \frac{U_{ref}}{2^{n-1}R} = \frac{U_{ref}}{2^{n-1}R} \sum_{i=0}^{n-1} b_i 2^i$$

$$U_{out} = I_S \cdot R_F = \frac{U_{ref} \cdot R_F}{2^{n-1}R} \sum_{i=0}^{n-1} b_i 2^i$$

Parallel conversion is only possible directly if the accuracy of the resistors (the resistors required are halved for the corresponding partial currents/voltages of the next higher bit) can be guaranteed. For example, for a 16-bit converter there are resistance tolerances for the most significant bit (MSB) of 0.0015%. In order to meet this requirement, the drift behavior must be the same for all resistors, especially when temperature fluctuations occur. It must also be taken into account that the switches also have a resistance that can lead to faulty converter results. Today's IC technologies enable an acceptable resistance ratio of 20:1, which means that a maximum 4–5 bits resolution can be generated error-free with this type of conversion. Several parallel converters are connected together to allow conversion of larger resolutions. For example, a 12-bit converter can be constructed from 3 blocks of 4-bit converters each. The blocks are coupled via current divider resistors in a ratio of 16:1. The partial currents are again added together with the help of an operational amplifier, as shown in Fig. 7.26.

7.7.2 R-2R Resistor Ladder Network

The R-2R resistor ladder (see Fig. 7.27) requires only two different resistance values as an alternative to the parallel converter. This has the advantage that these two

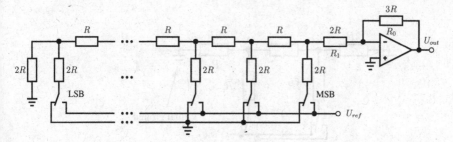

Fig. 7.27 Example circuit of an R-2R resistor ladder network

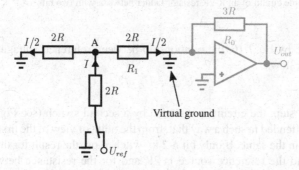

Fig. 7.28 Example circuit of an R-2R resistor ladder network with only one bit

resistors can be manufactured with very small tolerances. The two resistance values have a ratio of 2:1.

Thanks to the superposition principle of linear systems, the individual loops or sources can be considered separately. Assuming that the resistors and switches of the R-2-R chain conductor are ideal (linear), each switch can be considered separately, as shown in Fig. 7.28.

The operational amplifier is connected as an inverting amplifier, therefore the negative input can be considered as virtual ground. This simplification results in 3 resistors with only one resistance value 2R. With the help of the voltage divider rule, the voltage at point A is calculated as

$$U_A = U_{ref}/3 = I \cdot R \Rightarrow \frac{I}{2} = \frac{U_{ref}}{3 \cdot 2R} \qquad (7.16)$$

Since the input of the operational amplifier can be regarded as ideal, $I_{R_0} = I_{R_1}$. Due to the virtual ground, the output voltage at the operational amplifier is

$$U_{out} = U_{R_0} = \frac{I}{2} \cdot 3R$$

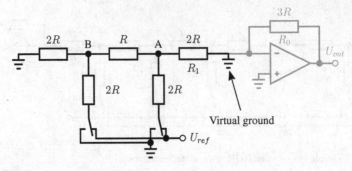

Fig. 7.29 Example circuit of an R-2R resistor ladder network with two bits

If $\frac{I}{2}$ is replace by Eq. 7.16, the relationship between reference voltage and output voltage is as follows.

$$U_{out} = U_{ref}/2$$

In the next step, the circuit is extended by a second switch (see Fig. 7.29). The resistors are extended in such a way that, from the point of view of the first switch, the resistors remain the same. If only bit n-2 is switched on, the result for the resistance between B and the reference source is 2R and for the resistance between B and virtual ground also 2R. If one combines the three resistors around A into one, the result again is the circuit in the illustration above with the voltage U_B at point B.

$$U_B = U_{ref}/3$$

If the resistor R_1 is brought together with the $2R$ resistor between point A and ground, the voltage U_A at point A is

$$U_A = U_B/2 = U_{ref}/6$$

Since the gain at the operational amplifier has remained the same, the ratio

$$U_{out} = U_{ref}/4$$

results. This procedure can be used for all the other bits.

7.7.3   Pulse Width Modulation (PWM)

One way to generate an analog signal is by pulse width modulation (PWM). This PWM signal is a *constant frequency* signal the *duty cycle* of which is changed depending on the voltage to be generated. This means that the longer the signal is at "1"

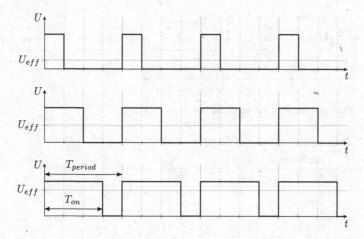

Fig. 7.30 Temporal course of various PWM signals

or "high" in each period, the higher the (averaged) output voltage (see Fig. 7.30). In an electrical circuit (e.g., microcontroller or DSP), a PWM signal is generated at the digital output. The digital output is followed by a low pass filter which smooths the PWM signal to create an analog average voltage $U_{off} = U_{ref} \cdot T_{on}/T_{period}$. So if you want to generate a voltage of $U_{off} = 2$ V at a signal level of $U_{ref} = 5$ V and a fixed period of $T_{period} = 1s$, the output must be switched to logical 1 within the period for T_{on} and then to logical 0.

$$T_{on} = \frac{U_{off}}{U_{ref}} \cdot T_{period} = \frac{2}{5} \cdot 1s = 0.4s$$

The PWM signal can also be transmitted directly to a motor driver, for example (see Sect. 9.2). In practice, the resolution of the clock/pause ratio is often not continuous, but quantized according to the resolution of the digital input value. For example, an 8-bit signal can only accept 256 different PWM output voltages.

7.7.4 Delta Sigma Converter

The delta-sigma converter works on a similar principle as the A/D converter of the same name (see Sect. 7.5.8). The input signal, consisting of n-bit digital values, is calculated with the maximum or minimum value based on the previous comparator result. This corresponds to the difference to the predicted value of the modulation. The value of the previous clock cycle is added to this result. This corresponds to the integration or summation. The comparator threshold is assumed to be 0. Since negative input values are encoded by the two's complement, it is sufficient to pass

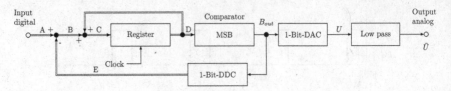

Fig. 7.31 Example circuit of a delta sigma converter

Table 7.2 Delta sigma conversion of the value -4 with $U_{ref} = \pm 7\,\text{V}$

	Initialization	Step 1	Step 2	Step 3	Step 4
A(n)	0	-4	-4	-4	-4
B(n)	0	-11	3	3	3
C(n)	0	-11	-8	-5	-2
D(n)	0	0	-11	-8	-5
E(n)	7	7	-7	-7	-7
B_{out}	0	0	1	1	1
U(n)	7V	7V	-7V	-7V	-7V

on the MSB. This encodes the sign. In the *1-Bit-Digital-to-Digital-Converter* (1-Bit-DDC) the highest achievable input value is accordingly output at $B_{out} = 0$, and the lowest achievable input value at $B_{out} = 1$. The generated bit stream as intermediate value is converted into analog voltages $\pm U_{ref}$ and smoothed by means of a low pass filter to obtain the continuous output signal.

A 4-bit converter is used as an example. This is shown in Fig. 7.31. As exemplary values, -4 is selected as number to be converted and the reference voltage as $U_{ref} = \pm 7\,\text{V}$

First (see Table 7.2), all values are initialized. In step 1, the additions are evaluated. Note, however, the time delay of one clock in the register, as a result of which the value is still 0 in position D. Hence

$$C(2) = B(2) + D(2) = B(2) + C(1) = -11 + 0.$$

In step 2, all effects are active for the first time, so that

$$B(3) = A(3) - E(2) = 3; \quad C(3) = B(3) + D(3) = 3 - 11 = -8$$
$$D(3) = C(2) = -11 \quad \text{bits}(3) = D(3) < 0? = 1$$

Accordingly, the filtered value $\overline{U}$ approaches the target end value with further steps.

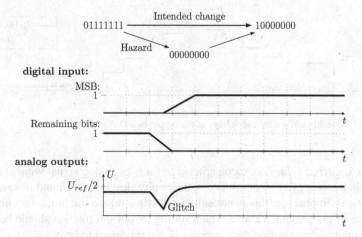

Fig. 7.32 Faulty edge of a digital signal

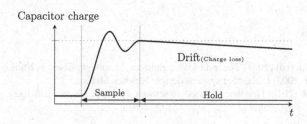

Fig. 7.33 Temporal course of a sample and hold step

7.8 Sample and Hold

Hazards can occur at the output of the D/A converter when changing the digital value. This means that if several bits flip at the same time, intermediate values can occur which can lead to voltage peaks or dips (see Fig. 7.32). These disturbances are amplified by an output amplifier which is often present. These errors can be avoided by using a (further) sample and hold circuit. This maintains the last stable value until a new stable state is reached in the next clock pulse. Maintenance of this value results in a discrete-valued but continuous-time signal (see Fig. 7.33). This step is not necessary in the case of pulse width modulation, since stable, continuous-time signals are already generated here.

7.9 Filters in the Reversal Process

At the output of the sample and hold circuit a stair-like voltage curve is created. These steps contain very high-frequency components. With the help of a low-pass filter, this staircase voltage can be smoothed (see Fig. 7.34).

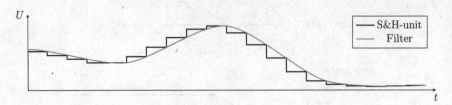

Fig. 7.34 Smoothing of a staircase voltage at the output of a D/A converter

If the digitized values are reconstructed to form an analog signal without digital manipulation (e.g., filter, controller ...), the two analog signals should theoretically be identical. In practice, this is not entirely possible due to the limited accuracy of the components presented above. The resulting loss of information should be taken into account when selecting the components and processes to be used.

References

1. Chen Wai Kai (ed) (2004) The electrical engineering handbook. Elsevier, Burlington
2. Heath Steve (2002) Embedded systems design. Newnes, Oxford
3. Fraden Jacob (2010) Handbook of modern sensors. Springer, Cham, Heidelberg

Chapter 8
Sensor Data Processing

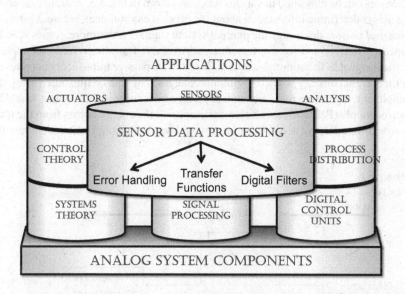

In this chapter, the basics of the acquisition of external stimuli in embedded systems by means of sensors are examined. First, possible measurement errors are categorized and the mathematical representation used for this is explained. In particular, the errors between the abstract sensor model and the real sensor are considered. Digital filters are then presented as an efficient way of processing signals in the processing unit. Finally, the special features of various specific sensor systems will be discussed.

8.1 Definitions of Sensor Technology

In the previous chapter, the general process of signal processing and the respective approaches were introduced. Recording of signals by means of sensors and their processing will now be examined more closely.

The term sensor comes from the Latin word *sensus*, which can be translated as *sense*. A sensor is a device for detecting, feeling or sensing physical and chemical quantities. In general, a sensor is a unit that receives and responds to a signal or stimulus. With a physical sensor, the output is an electrical signal such as a voltage, current or charge. Different values can be represented by amplitude, frequency or phase. A *stimulus* is a variable, property or condition that is perceived and converted into an electrical signal. In the measurement and control circuit, the sensor technology occupies the area of data acquisition between the transducer and the processing unit, as can be seen in Fig. 8.1.

Sensors can be classified in various ways, as shown in Fig. 8.3. *Extrinsic* or *external sensors* determine information about the process environment, whereas *intrinsic* or *internal sensors* determine the internal system status. Furthermore, sensors can be divided into active and passive transducers. An active transducer varies an applied electrical signal as the stimulus changes (Fig. 8.2). A passive transducer generates an electrical signal directly when the stimulus changes. This can be illustrated using the example of temperature measurement. A passive transducer in this context would be a thermocouple (Peltier/Seebeck element), since all the energy comes from the temperature difference. An active transducer would be a temperature-dependent resistor

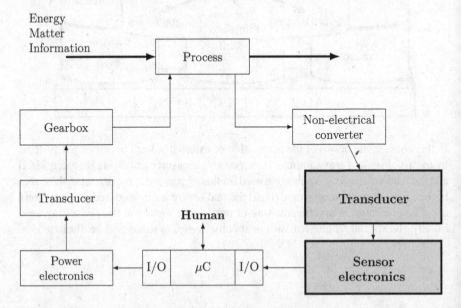

Fig. 8.1 Sensor data processing in the measurement and control cycle

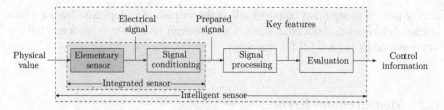

Fig. 8.2 Classification of sensors by level of integration

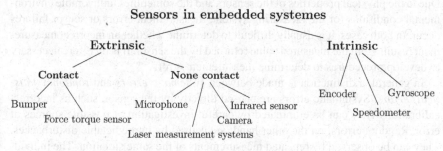

Fig. 8.3 Classification of frequently used sensors in embedded systems

(e.g., PT100), since an externally applied voltage is only varied in accordance with the temperature. Caution is advised with the term active/passive sensor. This term is used inconsistently and sometimes even in the opposite meaning throughout the literature. Depending on the context, it can further mean the transducer described above or the generation of the measured variable. In the latter case, the issue would be whether only passive incoming stimuli are measured (e.g., with a brightness sensor that only measures the ambient light present), or the change in an active measurement signal (e.g., a light barrier, where the change in brightness of a laser beam emitted by the sensor is measured).

For applications in the field of embedded systems, not only a physical sensor is used, but a sensor with corresponding sensor electronics. For example, the sensor electronics perform direct filter tasks, extract more complex information, digitize and scale. In order to take this into account in the classification, sensor systems are also differentiated by their various integration levels, as shown Fig. 8.2.

The *elementary sensor* picks up a measured variable and converts it directly into an electrical signal. The *integrated sensor* also contains a unit for signal processing, such as amplification, filtering, linearization and normalization. *Intelligent sensors* also contain a computer-controlled evaluation of the incoming signals. Often digital variables are already generated at the output which can be read directly by the embedded computer node. Typical outputs are binary values as are provided by e.g., light barriers and proximity switches, scalar outputs such as angle measurement generated by encoders, vector outputs such as the three forces and torques of a load cell or pattern recognition outputs such as the extraction of a person from a video stream. Information about other sensors and a deeper insight into sensor technology are provided by [1] and [2] for example.

8.2 Measurement Error

Due to the physical properties of the sensors and the sometimes unfavorable environmental conditions for the measuring system, measurement errors or sensor failures occur. In both cases, it is usually difficult to determine whether an incorrect measurement results from the measured value returned by the sensor. It is therefore necessary to develop procedures to determine measurement errors.

In general, a distinction is made between *systematic errors* and *random (statistical) errors*. Systematic errors are caused directly by the sensor, such as incorrect calibration. They can be eliminated by careful investigation of possible sources of error. Random errors, on the other hand, are caused by unpredictable disturbances. They can be observed by repeated measurements of the same situation. The individual measurements deviate from each other and usually fluctuate by an average value. This is shown as an example in Fig. 8.4 and Table 8.1.

The *arithmetic mean* is defined as

$$\bar{x} = \frac{1}{N} \sum_{i=1}^{N} x_i \tag{8.1}$$

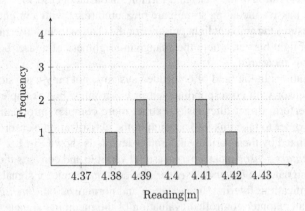

Fig. 8.4 Histogram of several consecutive distance measurements of the same situation

Table 8.1 Results of several consecutive distance measurements

4.40 m	4.40 m	4.38 m	4.41 m	4.42 m
4.39 m	4.40 m	4.39 m	4.40 m	4.41 m

For the above example this results to

$$\bar{x} = \frac{1}{10} \cdot (4.40\,m + 4.40\,m + 4.38\,m + 4.41\,m + 4.42\,m + 4.39\,m$$
$$+ 4.40\,m + 4.39\,m + 4.40\,m + 4.41\,m)$$
$$= \frac{1}{10} \cdot 44\,m = 4.40\,m$$

There are two forms for indicating the deviation of measurements.

- **Absolute error** Δx_i of a single measurement x_i is equal to the deviation from the mean value $\bar{x}$ of all N measurements $\{x_n | n \in \{1, \ldots, N\}\}$: $\Delta x_i = x_i - \bar{x}$.
- **Relative error** is the ratio of the absolute error to the measured value: $\frac{\Delta x_i}{x_i}$

However, these quantities are not meaningful when it comes to indicating the general accuracy of a measurement series or a sensor. Here the average error and the resulting standard deviation have become established. The latter can be used to determine the *measurement uncertainty*. The *average error* of the individual measurements is

$$\sigma_x^2 = \frac{1}{N-1} \sum_{i=1}^{N} (\Delta x_i)^2 = \frac{1}{N-1} \sum_{i=1}^{N} (x_i - \bar{x})^2 \qquad (8.2)$$

The formula contains the factor $N - 1$ and not N, since it is only divided by the number of comparison measurements.

This results in the following *mean error* (sample standard deviation)

$$\sigma_x = \sqrt{\frac{1}{N-1} \sum_{i=1}^{N} (x_i - \bar{x})^2} \qquad (8.3)$$

For the above example, this is

$$\sigma_x = \sqrt{\frac{1}{10-1} \cdot (0\,m^2 + 0\,m^2 + (-0.02\,m)^2 + (0.01\,m)^2 + (0.02\,m)^2}$$
$$\overline{+ (-0.01\,m)^2 + 0\,m^2 + (-0.01\,m)^2 + 0\,m^2 + (0.01\,m)^2)}$$
$$= 0.01154 m$$

The result of a measurement is usually written as

$$x = (\bar{x} \pm \sigma_x)\,[\text{Unit}] \qquad (8.4)$$

For the example, it is thus

$$x = (4.40\,m \pm 0.01154\,m)$$

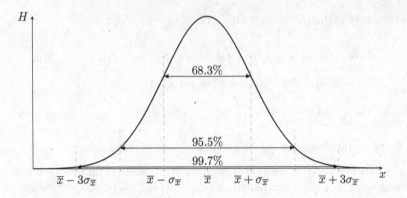

Fig. 8.5 Frequency H for an infinite number of measured values x with the standard deviation $\sigma_{\overline{x}}$ of the mean values of a Gaussian-distributed measurement

Specification of certain *confidence intervals* are also often found. These are the ranges in which the true value lies at a certain probability with an assumed normal distribution of the measurement series. For this purpose, the standard deviation of the mean values of the measurement series is first determined.

$$\sigma_{\overline{x}} = \sqrt{\frac{1}{N(N-1)} \sum_{i=1}^{N} (x_i - \overline{x})^2} = \frac{\Delta x}{\sqrt{N}} = \frac{\sigma_x}{\sqrt{N}} \tag{8.5}$$

The confidence interval of measurements can be specified with the help of a frequency distribution H of the measured values x with a normal distribution. In the case of large N, the confidence limit $\pm\sigma_{\overline{x}}$ indicates that the true value lies in the interval $\pm\sigma_{\overline{x}}$ around the mean value of all measurements with a probability of approx. 68%. If a confidence limit of 95% is required, the interval increases to $2 \cdot \sigma_{\overline{x}}$, at 99% to about $3 \cdot \sigma_{\overline{x}}$. This is shown in Fig. 8.5. Accordingly, the sample standard deviation can be used for the measured values themselves. Here, the confidence intervals indicate, that 68.3%/95.5%/99.7% of the measured values lie within the respective interval around the mean value of all measurements.

8.2.1 Error Propagation

If a derived value is calculated from several measured values, an uncertainty of measurement must also be specified. If the value to be calculated is $y = f(x_1, x_2, \ldots, x_N)$ and Δx_i the uncertainty of measurement of the individual values, the uncertainty of measurement Δy of the value to be calculated is

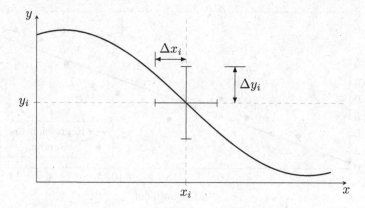

Fig. 8.6 Determined errors are shown as error bars at the measurement points

$$y + \Delta y = f(x_1 + \Delta x_1, x_2 + \Delta x_2, \dots, x_N + \Delta x_N)$$

$$\approx f(x_1, x_2, \dots, x_N) + \sum_{i=1}^{N} \frac{\partial f}{\partial x_i} \Delta x_i$$

$$\Delta y = \sum_{i=1}^{N} \frac{\partial f}{\partial x_i} \Delta x_i \text{ for } \Delta x << x_i \text{ (Taylor series)}$$

In the above equation, the partial derivatives represent weighting factors for error propagation. Examples of error propagation of a derived value are

- **Linear combination**

$$y = a_1 x_1 + a_2 x_2 + \dots + a_n x_n \quad \Delta y = \sum_{i=1}^{n} a_i \Delta x_i \tag{8.6}$$

- **Combined by multiplication**

$$y = a \cdot x_1 \cdot x_2 \cdot \dots \cdot x_n \quad \Delta y = y \cdot \sum_{i=1}^{n} \frac{\Delta x_i}{x_i} \tag{8.7}$$

The error variables determined in this way are usually shown in diagrams as error bars (see Fig. 8.6).

In general, the following rule of thumb can be used for error propagation:

- In an *addition* and *subtraction*, the *absolute* errors add up.
- In a *multiplication* and *division*, the *relative* errors add up.
- The difference between two quantities of almost equal size results in a large relative error, therefore it is better to measure the difference directly.
- Squaring doubles the relative error, taking the square root halves it.

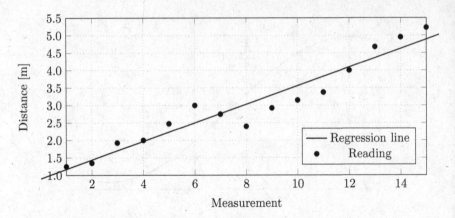

Fig. 8.7 Regression line of a measurement containing errors

Each measurement with a mean value $\overline{x}$ and standard deviation $\sigma_{\overline{x}}$ with large N is transformed into a Gaussian distribution (see Fig. 8.5).

$$f(x) = \frac{1}{\sigma_{\overline{x}}\sqrt{2\pi}}e^{-\frac{(x - \overline{x})^2}{2\sigma_{\overline{x}}^2}} \tag{8.8}$$

Often there is a relationship between two variables x and y, for example between the variables of current and voltage at a resistor. A particularly simple relation is the linearity of x and y with $y = m \cdot x + b$.

In this context, the coefficients of the linear equation can be determined by means of *linear regression* (see Fig. 8.7). Using linear regression, the two coefficients m and b in the case of an erroneous series of n measurements x_i and y_i can be determined as follows.

$$m = \frac{\sum_{i=1}^{n}(x_i - \overline{x})(y_i - \overline{y})}{\sum_{i=1}^{n}(x_i - \overline{x})^2} \text{ and } b = \overline{y} - m\overline{x} \tag{8.9}$$

where $\overline{x}$ and $\overline{y}$ are the mean values of the series of measured values.

For the quantitative determination of the linear dependence of two values x and y the empirical correlation coefficient r_{xy} is often given.

$$r_{xy} = \frac{\sum_{i=1}^{n}(x_i - \overline{x})(y_i - \overline{y})}{\sqrt{\sum_{i=1}^{n}(x_i - \overline{x})^2 \sum_{i=1}^{n}(y_i - \overline{y})^2}} \tag{8.10}$$

The closer r_{xy} is to 1, the stronger the linear dependence of the two values.

8.3   Transfer Function of Sensors

In the field of sensor technology we cannot necessarily assume a linear mapping of the input value to the output value. The concept of a transfer function is therefore somewhat more broadly defined here to $y = f(x)$. A distinction is made between the *ideal sensor transfer function*, which is based on a perfectly manufactured sensor without external disturbances, and the *real sensor transfer function*, which takes into account manufacturing tolerances, wear, environmental influences, etc. (see Fig. 8.8).

The relationship between stimulus and output signal is usually one-dimensional and linear.

The following applies to the linear transfer function: $y = a + b \cdot x$. The gradient b is also referred to as *sensitivity*. It is a measure of the distinctness of the output values for small differences in the corresponding input values.

Further important transfer functions of sensors are e.g.,:

- **Logarithmic** transfer function: $y = a + k \cdot \ln x$, k=constant
- **Exponential** transfer function: $y = a \cdot e^{kx}$
- **Polynomial** transfer functions: $y = a_0 \cdot x^0 + a_1 \cdot x^1 \ldots$

The sensitivity for non-linear transfer functions is defined as follows for each input value x_{in}.

$$b = \frac{dy(x_{in})}{dx} \tag{8.11}$$

The dynamic range of a stimulus that is detected by a sensor is called the *span* or the *full scale input*. This is the lowest and highest stimulus value allowed for a sensor. Stimuli outside the full scale input can damage the sensor.

The *full scale output* of a sensor is the interval between the output signal with the smallest and largest (permissible) stimulus applied.

The *accuracy* of a sensor describes the maximum deviation between the ideal values and the values output by the sensor. As with any measurement, reference is made to systematic and random errors of a sensor.

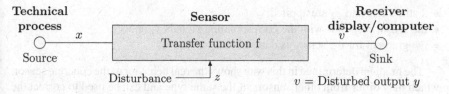

Fig. 8.8 Transfer function of a sensors

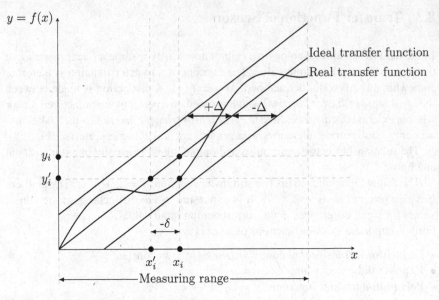

Fig. 8.9 Comparison of the real and the ideal transfer function of a sensor

8.3.1   Properties of Sensor Transfer Functions

The ideal transfer function is designated by $y = f_{\text{ideal}}(x)$ and the real transfer function by $y' = f_{\text{real}}(x)$. If the ideal transfer function is used to map the result y_i' to the stimulus, we get x_i' and $\delta = x_i - x_i'$ (see Fig. 8.9).

Sensors often have systematic errors. The sensor output is shifted by a constant value for each stimulus. However, this error is not necessarily evenly distributed over the input area. To avoid this, sensors are calibrated after manufacture.

The displacement of the sensor output can be different for each manufactured sensor due to the material. The gradient for each sensor is therefore determined for calibration purposes:

- 2 stimuli x_1 and x_2 are applied,
- the sensor responds with the corresponding signals y_1 and y_2,
- the gradient for the sensor is determined.

The gradient determined in this way shows the real behavior of the concrete sensor, which may differ from other sensors of the same type and can be used to correct the measured values. This results in the problem that the gradient will not correspond to the real gradient due to measurement errors (see Fig. 8.10). Some possible error types are listed below.

A *hysteresis error* is the different deviation of the output signal of a sensor for a certain stimulus value, depending on the direction from which the stimulus approaches this value. A hysteresis is shown in Fig. 8.11.

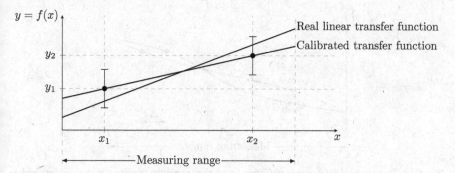

Fig. 8.10 Erroneous calibration with a linear transfer function under influence of errors

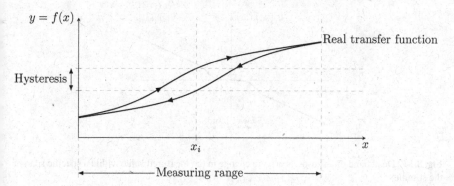

Fig. 8.11 Hysteresis of a transfer function

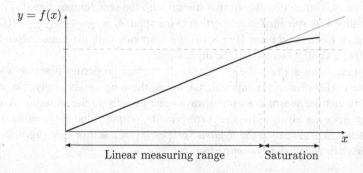

Fig. 8.12 In the saturation region, the sensor transfer function deviates from previously linear course

Almost every sensor has working range limits and many sensors have a linear transfer function, but above a certain stimulus value, the output required is no longer generated. In this case, the term *saturation* is used (see Fig. 8.12).

A sensor can produce different output values under the same conditions. This error corresponds to the *repeatability*, which is shown in Fig. 8.13. For two calibration

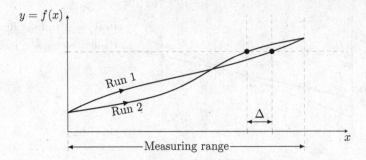

Fig. 8.13 Repeatability of a sensor, different measurement results from two measurements under the same conditions

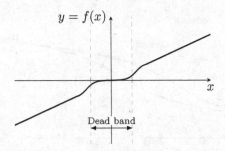

Fig. 8.14 Dead band of a sensor, nearly no change in the measured value within a specific span of the stimulus

cycles, the maximum distance of two stimuli with the same output signal is Δ. The repeatability δ_r is specified in proportion to the span. $\delta_r = \frac{\Delta}{\text{Measurement area}} \cdot 100$.

A sensor has a *dead band* (Fig. 8.14) if it responds with the same output signal (often 0) in a contiguous area of the input signal.

For static input signals, the previously mentioned properties describe a sensor completely. However, if the input signal varies, these no longer apply. The reason for this is that the sensor does not always react directly to the stimulus. A sensor therefore does not always give the corresponding output value at the same time as the stimulus. This is called the *dynamic properties* of a sensor (see Fig. 8.15). The resulting error is called a *dynamic error*.

8.4   Digital Processing

The analog signals coming from elementary sensors are converted into digital signals for further processing, e.g., in a microcontroller. However, these signals are often still faulty and must be filtered. The methods used for this purpose are presented below.

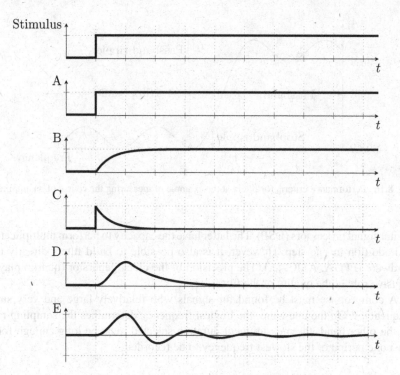

Fig. 8.15 Examples of different dynamic reactions of sensors stimulated with the unit jump

Although analog filters are still used, for example, in the high-frequency range, filtering of signals that have already been digitized has largely become established in the area of embedded systems. This is due, on the one hand, to the fact that no additional components are required, which can also be additional sources of interference, and on the other hand to the fact that these filters can be changed comparatively simply during runtime and adapted to the ambient conditions. For this purpose, these digital filters are simply programmed on digital computers. Signals are processed as (value and time) discrete number sequences. Digital filters have similar characteristics to analog filters. This means that the transition from passband to stopband depends to a large extent on the order of the filter (see Fig. 8.16). Digital filters have the advantage that the order of the filter is limited only by the computing power of the processor. Filters with an order greater than one hundred are feasible.

Since digital filters work with sampled signals, the maximum frequency of the filter is limited on the one hand by the speed of the A/D converter and on the other hand by the clock frequency of the filter. The clock frequency of the filter indicates how often the filter is accessed. If the sampling frequency is reduced, filters can be built for "very slow" signals, which are very difficult to realize with analog filters.

The calculation of a filter step requires a very large number of multiplications and additions. Therefore, digital filters are used on microcontrollers (μC) or mostly on

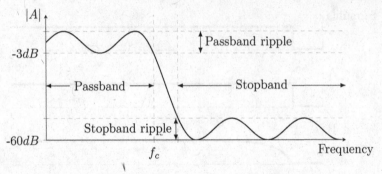

Fig. 8.16 Performance criteria for filters at the example of measuring the course of an amplitude

digital signal processors (DSP). The latter have the capacity to perform multiplication and addition in one step. However, it is also possible to build filters directly into hardware (FPGA, ASIC, ...). The precision of the clock generator (quartz) plays a decisive role in the quality of the filter.

A compromise must be found for signals with relatively large and very small frequencies. On the one hand, the highest frequency determines the sampling rate, on the other hand, the measurement duration selected must be long enough for at least one period of the slowest frequency to be recorded.

8.4.1   FIR Filter

FIR (*finite impulse response*) filters calculate the filtered value from a finite number of values in the input sequence $x(k)$. The number of values considered indicates the order of the filter. The higher the order of an FIR filter, the steeper the edge can become. An example circuit of an FIR filter is shown in Fig. 8.17. The difference equation for such a filter is

$$y(n) = \sum_{m=0}^{N} a_m \cdot x(n - m) \tag{8.12}$$

One advantage of FIR filters is that they can never become unstable because of their finite impulse response. Another advantage is that they can have a constant group delay. The group delay describes the difference between the times that different frequencies take to pass through the filter. FIR-Filter with this feature are called *linear phase*. Signals are not distorted by a constant group delay. A disadvantage of FIR filters is the high number of weights required to achieve good filter properties. Bandpass and band rejection filters are not feasible with an order smaller than 10 and therefore require many resources as FIR filters.

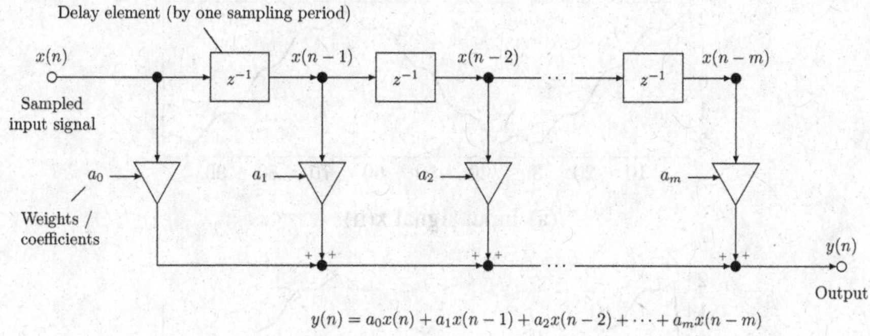

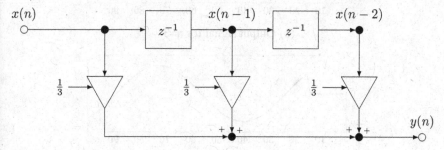

Delay element (by one sampling period)

$y(n) = a_0 x(n) + a_1 x(n-1) + a_2 x(n-2) + \cdots + a_m x(n-m)$

Fig. 8.17 Principle structure of an FIR filter

Fig. 8.18 Averaging filter with $N = 3$ coefficients

The best-known FIR filter is the averaging filter (see Fig. 8.18). It adds up the last N values and divides them by the number of values, thus acting as a low-pass. Considered as a mathematical sequence, this gives

$$y(n) = \frac{1}{N} \cdot \sum_{k=0}^{N-1} x(n-k)$$

In the case of $N = 3$, the function would therefore be

$$y(n) = \frac{1}{3} \cdot (x(n) + x(n-1) + x(n-2))$$

The filter would correspond to Fig. 8.18.

If the input signal $x(n) = 2 \cdot cos(n) + \frac{1}{5} \cdot cos(8n)$ is applied to this filter (see Fig. 8.19a), the output signal would correspond to Fig. 8.19b.

You can see that the higher-frequency signal component is already damped, but only slightly. If the filter was created with n = 15 instead, the output signal would be as in Fig. 8.19c. Here the high-frequency signal component is practically no longer visible. However, the problem that digital filters need some time to tune in to the signal

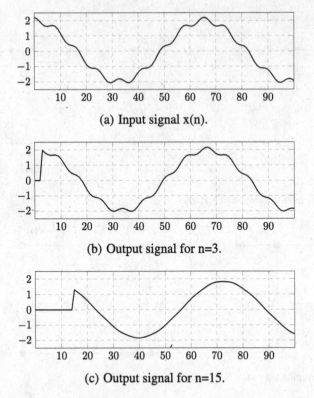

(a) Input signal x(n).

(b) Output signal for n=3.

(c) Output signal for n=15.

Fig. 8.19 Example of an averaging filter with the input function $x(n) = 2 \cdot cos(n) + \frac{1}{5} \cdot cos(8n)$

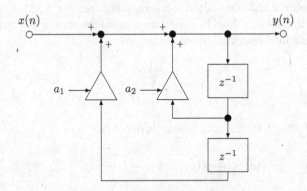

Fig. 8.20 Example of an IIR filter

clearly emerges here. This time is called the *transient phase*. A similar phenomenon occurs at the end of the process. Since the filter still calculates values for n steps after the end of the signal, the signal slowly decays here, which is thus referred to as the *decay phase*.

8.4.2 IIR Filter

IIR (*infinite impulse response*) filters have an infinite impulse response. Because the feedback of an IIR filter means that all previous values of the input sequence are included in the result, the filter is five to ten times more effective than a FIR filter of comparable order. This is particularly noticeable when the number of coefficients is limited by the hardware. Due to the lower order, the transit time of the signal components through the filter is lower. Due to the feedback of the output sequence, the impulse response theoretically decays only after an infinite time. If the filter parameters are selected incorrectly, the filter may even oscillate and become unstable. The order of an IIR filter is determined by the recursive (M) and non-recursive (N) parts. The higher degree determines the order of the filter. An example of an IIR filter is shown in Fig. 8.20. The difference equation is generally

$$y(k) = \sum_{i=0}^{N} b(i) \cdot x_{k-i} + \sum_{j=1}^{M} a(j) \cdot y_{k-j} \tag{8.13}$$

For the system in Fig. 8.20 the result is

$$y(k) = x(k) + a_1 y(k-1) + a_2 y(k-2)$$

Due to the dependence on earlier output values, it is easily imaginable that the wrong choice of a_1 or a_2 can result in echoing and unstable behavior

A disadvantage is that the group delay is no longer constant due to recursion. As a result of the different group delays, individual frequency components need longer to pass through the filter. The result is a distorted signal at the output. IIR filters are therefore difficult to use in audio equipment, for example.

8.4.3 Examples of Digital Filters

The design of filters, especially those of a higher order, is so complex that it is virtually impossible to carry out without computer support. However, objectives and resources play a major role in filter design, which is why various design methods have emerged, some of which are presented below on the basis of their prioritized criteria. A graphical overview can be seen in Fig. 8.21.

A popular filter model is the *Chebyshev* filter. This type of filter is designed to achieve the steepest possible drop in the frequency response beyond the cut-off frequency. They also have very good damping in the stopband. This is achieved with a non-monotonous, oscillating course in the passband. In addition, these filters have a comparatively long transient phase.

Bessel filter, on the other hand, oscillate very cleanly and have an almost linear transfer function in the passband, while the group delay is also very constant. How-

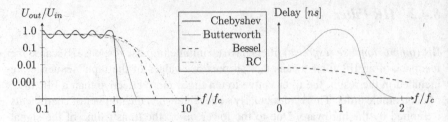

Fig. 8.21 Amplitude course and delay of different filters

ever, the drop beyond the cut-off frequency is not very steep, which has a negative effect on damping in the stopband.

A good compromise is the *Butterworth* filter. The amplitude in the passband is very flat and it vibrates quite fast. Its damping lies between the other two filters. The RC low-pass (see Sect. 3.5.1) is a simple 1st order Butterworth filter.

8.5   Internal (Intrinsic) Sensors

As described in the introduction, intrinsic sensors are used to record the internal states of a system. Typical intrinsic sensors include position sensors, velocity sensors, acceleration sensors, gyroscopes or geomagnetic sensors, which can be designed as optical encoders, magnetic inductive encoders and potentiometers. Extrinsic sensors are, for example, ultrasonic sensors or camera systems. Some important measuring principles will be presented below using optical encoders and capacitive acceleration sensors as examples.

8.5.1   Optical Encoders

Optical encoders are used to measure distances and speeds, in particular to measure the movement of axes. They can be designed as relative or absolute encoders, as well as continuous or incremental. For a relative, incremental measurement, two light sources arranged at a certain distance are installed above a black-and-white grating. This is shown in Fig. 8.22. The black-and-white grid or the sensor can either move in a linear direction or rotate. The black, opaque and white, translucent surfaces have the same width. On the opposite side of the two light sources there are two detectors (e.g., photodiodes), which deliver a current proportional to the measured light intensity. The doubling of the light source-light detector pair serves to determine the direction, as explained below. The movement of the black and white grid between light source and detector generates an almost rectangular signal. If a certain brightness threshold is defined above which a one and below which a zero is to be returned at the detector

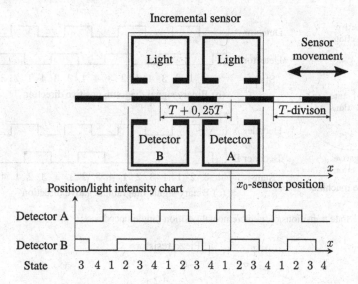

Fig. 8.22 Principle of a relative, incremental position-/angle-sensor

Table 8.2 States of an incremental position-/angle-encoder

States	1	2	3	4
Detector A	1	1	0	0
Detector B	0	1	1	0

output, this can be used as a digital signal. The number of transitions from zero to one or from one to zero is proportional to the linear or rotational movement of the black-and-white grid.

If the two detectors are set apart by $nT + 0.25T$, the direction of the movement can also be determined. Depending on the direction of rotation, there is a clear state transition (1-2-3-4 or 4-3-2-1, see Table 8.2), which means that the direction of movement can be clearly determined (see Fig. 8.23). Since encoders directly return a digital counter value, no A/D conversion is necessary.

8.5.2 Acceleration Sensors

In addition to measuring the acceleration of an embedded system, acceleration sensors also serve as a basis for calculating its location and position. They can therefore be found in almost every mobile device. Various technologies are used for this purpose. Examples of mechanical acceleration sensors include piezoresistive sensors (see Fig. 8.24), piezoelectric sensors, Hall effect sensors, thermal sensors and capac-

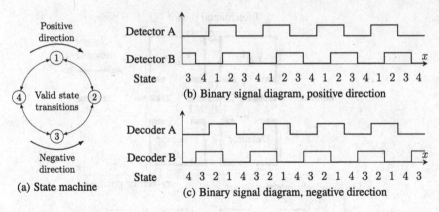

Fig. 8.23 State transitions of an incremental position-/angle-encoder

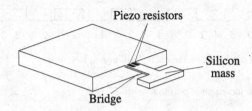

Fig. 8.24 Piezoresistive acceleration sensor

itive sensors. Examples of other acceleration sensors are resonant sensors and fiber optic sensors.

A capacitive acceleration sensor is presented below which can be found in many embedded systems (e.g., smartphones) and which can be used to explain the non-electrical and electrical conversion of measured variables in the measurement and control cycle. To measure an acceleration, the basic equation of mechanics is used, in which a relationship between force, acceleration and mass is described. If a mass is attached to a spring element, as in Fig. 8.25, with $F = k \cdot y$ (k spring constant, y deflection), the acceleration can be calculated from $F = m \cdot a$ to

$$a = \frac{k \cdot y}{m} \tag{8.14}$$

If this mass, which is bounded by two conductive plates, is embedded between two further plates C_1 and C_2, a differential capacitor is formed, as can be seen in Fig. 8.25. The plates with the distance d_0 are charged until the potential difference of the two plates is equal to the applied voltage U. There is an approximately homogeneous field between the plates with field strength $E = \frac{U}{d_0} = \frac{\sigma}{\varepsilon_0}$, where $\sigma = \frac{Q}{A}$ is the charge density, A the plate area and ε_0 the dielectric constant. This results in a capacity for the capacitor of

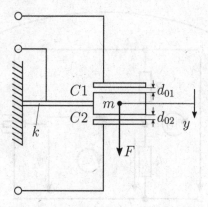

Fig. 8.25 Structure of a differential capacitor

$$C = \frac{Q}{U} = \frac{Q \cdot \varepsilon_0}{\sigma \cdot d_0} = \frac{Q \cdot \varepsilon_0 \cdot A}{Q \cdot d_0} = \frac{\varepsilon_0 \cdot A}{d_0} \tag{8.15}$$

Equation 8.15 shows that the capacity is inversely proportional to the plate spacing.

The differential capacitor in Fig. 8.26 will now be considered. The distances between the plates d_{01} and d_{02} change during deflection y by $\pm\Delta d$. This results in changes of the capacities C_1 and C_2.

Due to the acceleration of the mass, the plate distances d_{01} and d_{02} change by $\pm\Delta d$.

Since the capacitance of a capacitor depends on the spacing and the plate surface area, the capacitances C_1 and C_2 change in opposite directions. The bridge circuit shown in Fig. 8.26 is suitable for determining the capacitance changes and for indirectly calculating the movement of the mass.

The measuring voltage is U_m and U_s an AC voltage. Here we exploit the fact that a capacitor behaves like a complex resistor with the impedance $Z_C = \frac{1}{j\omega C}$ when an AC voltage is applied (see Sect. 3.4.4). From the mesh rule and the voltage divider rule, it follows

$$M_1 : U_m = U_1 - U_2; \quad M_2 : U_2 = U_s \frac{R}{R+R} = \frac{1}{2}U_s$$

$$M_3 : U_1 = U_s \frac{\frac{1}{j\omega C_2}}{\frac{1}{j\omega C_1} + \frac{1}{j\omega C_2}} = U_s \frac{C_1}{C_1 + C_2}$$

By means of the dependence of the capacitance on the deflection of the mass shown in Eq. 8.15, the measuring voltage U_m can now be determined as follows as a function of the mass deflection.

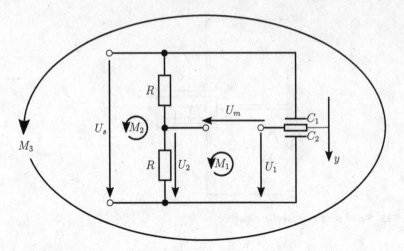

Fig. 8.26 Bridge circuit for a differential capacitor

$$U_m = U_1 - U_2 = U_s \left(\frac{C_1}{C_2 + C_1} - \frac{1}{2} \right) = \frac{U_s}{2} \left(2\frac{C_1}{C_2 + C_1} - 1 \right) = \frac{U_s}{2} \left(\frac{-C_1 - C_2 + 2C_1}{C_2 + C_1} \right)$$
$$= \frac{U_s}{2} \left(\frac{C_1 - C_2}{C_2 + C_1} \right)$$

By using $C_1 = \frac{\varepsilon_0 \cdot A}{d_0 + \Delta d}$ and $C_2 = \frac{\varepsilon_0 \cdot A}{d_0 - \Delta d}$ with $d_{01} = d_0 + \Delta d$ and $d_{02} = d_0 - \Delta d$ in the equation above, the following result is obtained

$$U_m = -U_s \frac{\Delta d}{2d_0} \tag{8.17}$$

$$\Rightarrow \Delta d = -\frac{2d_0}{U_s} U_m \tag{8.18}$$

This means, that the change in distance between the two capacitors can now be determined by the measuring voltage U_m, the AC voltage U_s and the initial distance between the plates d_0. The acceleration a can now be determined from the change in distance (deflection of the seismic mass) using Eq. 8.14. $y = \Delta d$ is used for this purpose. This results in a proportional relationship between the measured voltage U_m and the acceleration a.

$$a = -\frac{2d_0 k}{U_s m} \cdot U_m \tag{8.19}$$

In this example, the non-electrical measurement conversion described in the measurement and control circuit is the determination of the acceleration by measurement of the deflection of the mass. Since the displacement of the mass is proportional to the capacitance of the capacitor, this displacement can be obtained by determining voltages at the differential capacitor. This is an electrical transducer. This sensor can be constructed as a micromechanical system and can therefore be produced very cost-effectively.

8.6 External (Extrinsic) Sensors

Many embedded systems are used to capture environmental properties such as temperature or distance. Using the example of the forklift truck, localization of the vehicle within its working area is relevant. Detection of obstacles such as people is a consideration. Such systems are equipped with extrinsic sensors for this purpose. Extrinsic sensors include proximity, distance, position, tactile and visual sensors. The class of tactile sensors includes touch, sliding and force-torque sensors, proximity sensors are divided into inductive, capacitive, optical and acoustic sensors. Distance sensors can be optical, acoustic and radar sensors. 3D sensors, CCD and photodiodes are visual sensors and ground based radio systems, natural/artificial landmark detection and (differential) GPS are position sensors.

8.6.1 Strain Gauges

External forces that are in equilibrium (otherwise acceleration!) and act on a solid body lead to changes in the shape and volume of that body. If these forces cease, the change in shape and volume is completely reduced. The prerequisite for this is that the deformation does not exceed a certain limit. A body with this property behaves *elastically*. Such deformations can be used to measure the applied force or the resulting strain via a change in resistance.

The characteristic parameter for the stress on solid particles is the mechanical stress $S = \frac{F}{A}$. The force F is broken down into a normal component F_n and a tangential component F_t for easier calculation (see Fig. 8.27).

- **Normal stress** $\sigma = \frac{F_n}{A}$
- **Shear stress** $\tau = \frac{F_t}{A}$ (Tangential stress)

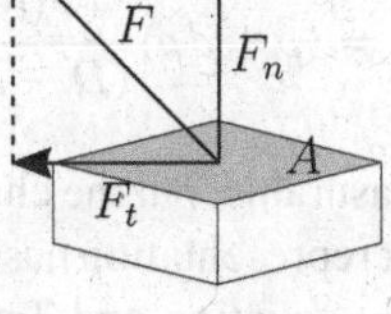

Fig. 8.27 Braking down the applied force F in a normal component F_n and a tangential component F_t

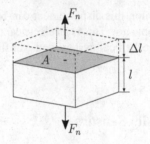

Fig. 8.28 Elongation of a body in the elastic range

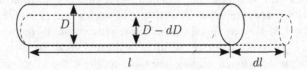

Fig. 8.29 Change of length and diameter of a body in the elastic range

The prerequisite is that the forces are evenly distributed over the surface area A. This is assumed to be given if the area is sufficiently small. The normal force F_n (respectively, the mechanical normal stress σ) causes a change in length Δl of the body (see Fig. 8.28). In the elastic range, this extension Δl is proportional to F_n (or σ): $\sigma \sim \frac{\Delta l}{l}$.

The relative change in length is defined as elongation $\varepsilon = \frac{\Delta l}{l}$ (unit is usually [μm/m]). *Hooke's law on elastic deformation* follows from this: $\sigma = E \cdot \varepsilon$. The proportionality factor E is called *Young's modulus* (material constant, unit [N/mm^2]).

The electrical resistance of a wire changes under the influence of elongation ε (2.4). This effect is exploited by strain gauges. Two factors play a role here. The increase of the length l of the wire to the length $l + dl$ and the reduction of the diameter D by the amount dD (see Fig. 8.29). The magnitude of the ratio of relative diameter change to elongation ε is called *Poisson's ratio* $\mu = \frac{dD/D}{\varepsilon}$.

Thus, the resistance of the unloaded strain gauge is

$$R = \rho \frac{l}{A} = \rho \frac{4 \cdot l}{D^2 \pi}$$

For the resistance of the loaded strain gauge, this results to

$$R + dR = (\rho + d\rho) \frac{4 \cdot (l + dl)}{(D - dD)^2 \pi}$$

The change in resistance is measurable, but the change in the influencing variables is unknown. Therefore, a different representation has become established here, which results from transformation of the equation and Taylor series development for the relative resistance change $\frac{dR}{R}$.

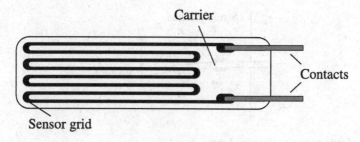

Fig. 8.30 Example of an industrial strain gauge

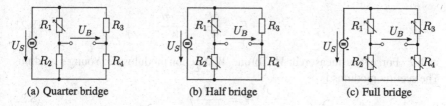

(a) Quarter bridge (b) Half bridge (c) Full bridge

Fig. 8.31 Wheatstone bridges for strain gauges

$$\frac{dR}{R} = k\frac{dl}{l} = k \cdot \varepsilon \tag{8.20}$$

The resistance change is proportional to the elongation, the proportionality factor k (called k-*factor*) describes the sensitivity of the strain gauge and depends on the material.

Industrial strain gauges have a k factor of approximately 2. The measuring grid usually consists of meander-etc.hed Constantan foil in order to increase the overall length and thus the sensitivity. Changes in resistance are caused by elongation or compression in the horizontal direction, transverse contraction (elongation or compression in the vertical direction) is assumed to be negligible. An example is shown in Fig. 8.30.

In order to measure these changes effectively, resistance bridges are used, as with the capacitive acceleration sensor. *Wheatstone bridges* distinguish between active and passive bridge branches. Typical resistance bridges for strain gauges are quarter bridges, half bridges and full bridges, as shown in Fig. 8.31. Quarter bridges consist of one active branch (e.g., strain gauge) and three passive branches (fixed resistors). With half bridges, strain gauges are attached to the material in such a way that R_1 is stretched and R_2 compressed, with full bridges in such a way that R_1 and R_4 are stretched and R_2 and R_3 are compressed.

As an example, strain gauges on the bending beam will be considered, as shown in Fig. 8.32. Strain gauges are attached to the top and bottom of a bending beam (width b, height h) at a distance a from the point where the force is applied. This leads to alternating elongation and compression of the strain gauges (half bridge).

The bending stress on the surface of the beam is

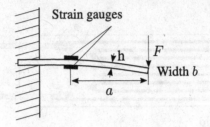

Fig. 8.32 Strain gauges at a bending beam

$$\sigma = \varepsilon \cdot E = \frac{M}{W} = \frac{F \cdot a}{W} \tag{8.21}$$

with F: Force to be measured, M: Torque, W: Section modulus, E: Young's modulus. The section modulus is

$$W = \frac{b \cdot h^2}{6} \tag{8.22}$$

This results in the elongation of the bending beam at the location l.

$$\varepsilon = \frac{a \cdot F}{W \cdot E}$$
$$= \frac{6 \cdot a \cdot F}{b \cdot h^2 \cdot E}$$

If the elongation determined in this way is used in Eq. 8.20, the change in resistance of the respective strain gauges results.

$$\frac{dR}{R} = k \cdot \frac{6 \cdot a \cdot F}{b \cdot h^2 \cdot E} \tag{8.23}$$

Since a, b, h and E are constant, the change of resistance is proportional to the applied force.

$$\frac{dR}{R} \sim F \tag{8.24}$$

8.6.2   Force-Torque Sensors

Load cells are used to record forces and torques, such as those occurring between the robot and the workpiece. They consist of a combination of several force transducers, such as strain gauges or piezo crystals.

The "spoked wheel shape" has emerged as a typical form for load cells with strain gauges (see Fig. 8.33b). The approach is that acting forces change the length of the elastic bars. Strain gauges are mounted on these bars and their resistance changes as

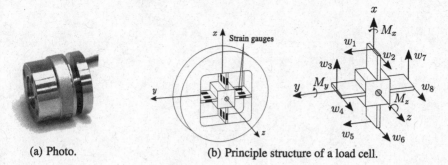

(a) Photo.   (b) Principle structure of a load cell.

Fig. 8.33 Load cell

a result. The resulting system of equations for the strains and the resulting forces in all spatial directions can be calculated using the *decoupling matrix* (Eq. 8.25), where the constants k_{ij} indicate the contribution of the strain gauge w_i to the force or torque j. It is

$$
\begin{pmatrix} F_x \\ F_y \\ F_z \\ M_x \\ M_y \\ M_z \end{pmatrix} = \begin{pmatrix} 0 & 0 & k_{13} & 0 & 0 & 0 & k_{17} & 0 \\ k_{21} & 0 & 0 & 0 & 0 & k_{26} & 0 & 0 \\ 0 & k_{32} & 0 & k_{34} & 0 & k_{36} & 0 & k_{38} \\ 0 & 0 & 0 & k_{44} & 0 & 0 & 0 & k_{48} \\ 0 & k_{52} & 0 & 0 & 0 & k_{56} & 0 & 0 \\ k_{61} & 0 & k_{63} & 0 & k_{65} & 0 & k_{67} & 0 \end{pmatrix} \cdot \begin{pmatrix} w_1 \\ w_2 \\ w_3 \\ w_4 \\ w_5 \\ w_6 \\ w_7 \\ w_8 \end{pmatrix} \qquad (8.25)
$$

These load cells can be used, for example, to measure weights or torques, whereby the measuring range can be between 0.5 N and several tens of MN depending on the specific sensor. They are used, for example, in excavators and robot arms for load measurement and in electric scales.

References

1. Lee, Edward Ashford and Sanjit A Seshia: *Introduction to embedded systems: A cyber-physical systems approach*. Mit Press, Cambridge, 2016
2. Fraden Jacob (2010) Handbook of modern sensors. Springer, Cham, Heidelberg

Chapter 9
Actuators

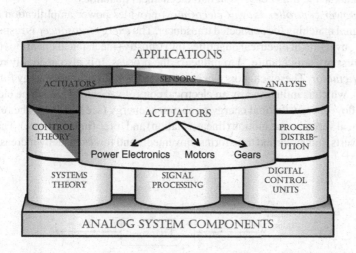

This chapter introduces various actuators that embedded systems can use to affect their environment physically. In addition to the electric motors, represented by the DC motor and the stepper motor, various gear types will also be presented. Gears can be used to adjust the forces or torques and the (rotational) speeds of the actuators. Finally, rheological fluids, piezoelectric actuators and various thermal actuators are introduced as alternative actuators. In addition to the basic principles, the use of these components is also explained using examples.

9.1 Actuator Components

Most reactive embedded systems aim to influence their condition or environment based on the measured sensor values and the conclusions drawn from them. For example, the purpose of a forklift robot is to pick up objects in its environment and transport them to a given destination. In doing so, it must also change its own position and avoid obstacles. Therefore, the previously generated digital control signal of the processing unit must be used to generate an analog, electrical manipulated variable and finally a mechanical movement (see Fig. 9.1). This happens in three steps. First, the D/A converted signal is amplified in order to adapt it to the power range of the actuators. Then the actual conversion of electrical power into mechanical power takes place. For example, an electric motor converts electrical voltage and current into speed and torque at the shaft of the motor. Finally, a gear or a displacement transducer is used to adjust the speed or torque. Various principles of action can be used to enable the final conversion into mechanical quantities.

The *power controller* (*power electronics*) provides power amplification and an input signal matching the connected transducer. The *energy transducer* is responsible for the conversion of electrical quantities into (mostly) mechanical quantities (movement, pressure, temperature...) to influence the process. It is also generally referred to as an *actuator*. These actuators can be classified by their primary energy form (see Fig. 9.3), which is influenced by an electrical control variable. This can be electrical energy, flow energy, thermal energy or chemical energy (see Fig. 9.2). The resulting mechanical variables are motion (linear, rotation) and force (linear, torque). The *gearbox* converts, transmits and transforms movements and forces (usually increasing the travel).

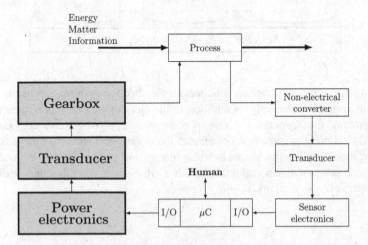

Fig. 9.1 Actuators in the measurement and control cycle

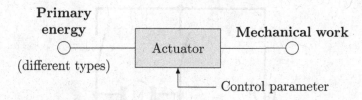

Fig. 9.2 Effect principle of actuators

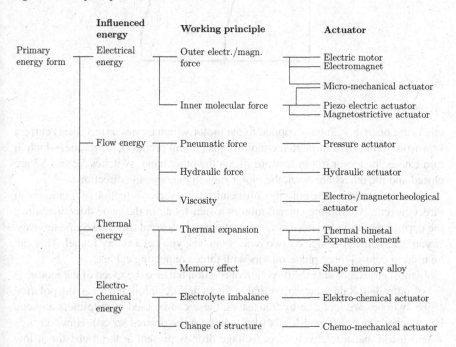

Fig. 9.3 Classification of actuators

9.2   Power Electronics

Usually, electric motors cannot be controlled directly by an embedded processing unit (e.g., microprocessor), as these do not supply sufficiently large currents for standard motors. This requires electronics that additionally convert the calculated digital variables into analog variables such as voltage and current. The most commonly used technique is an *H bridge* in combination with a pulse width modulated signal (PWM) (see Sect. 7.7.3). The designation "H bridge" comes from the H-shaped arrangement of the elements in the circuit diagram, as can be seen in Fig. 9.4.

The elements marked $S1$ to $S4$ represent switches. They can take the form of transistors, thyristors or relays. This H-shaped arrangement is used to reverse the polarity of the motor and to switch it off. If switches $S1$ and $S4$ are closed and the

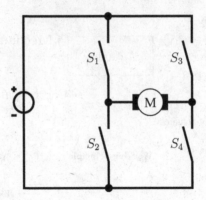

Fig. 9.4 Circuit of an H bridge

others are open, a voltage is applied to the motor which causes a (technical) current flow from left to right through the motor. This current produces the torque, which in turn causes the motor to run forward. If, on the other hand, switches $S2$ and $S3$ are closed and the others are open, the motor runs in the opposite direction.

However, reversing the polarity only results in a reversed direction of rotation with direct current. Alternating current motors would rotate in the same direction, since the current direction of alternating current changes several times a second anyway. If you extend the H-bridge by two more switches, you get a rotary funnel. This can be used to control three-phase motors with three connecting cables.

In most cases, not only the direction of rotation but also the speed of the motors is to be controlled. Relays are only suitable for switching on/off and reversing polarity. If the switches are replaced by transistors, these can be used as amplifiers and can be switched on at different levels (depending on the desired speed). However, this would mean that a relatively large voltage drop is present at the transistor at low speeds. This voltage drop causes a power loss.

$$P_L = U_{\text{Voltage drop}} \cdot I_{\text{Motor}} \tag{9.1}$$

The transistor converts this power into heat. The power dissipation can even be greater than the effective power. Pulse width modulation (PWM) is an energetically more effective method. A voltage value proportional to the analog input variable is simulated by periodically switching the supply voltage on and off with different duty cycles. Because the transistors are either de-energized when switched off or the voltage drop is (almost) zero when switched on, the power dissipation is also (almost) zero. The voltage applied to the motor can be calculated from the PWM's duty cycle.

$$U_{Motor} = U \cdot \frac{t_{on}}{t_{Period}} \tag{9.2}$$

Due to the inertia of the motor or the capacitances and inductivities in its design, the PWM input signal is smoothed to such an extent that the motor runs at an (approximately) constant speed, which is slower than the maximum speed by the clock-pause ratio.

9.3 Electric Drives

Electric motors are the best known form of actuators for embedded systems. They can be divided into linear and rotary drives. Rotary electric motors are again subdivided into DC motors, AC motors and stepper motors. Strictly speaking, piezoelectric actuators would also fall into this category, but due to their different functional principle they are treated separately.

Electric motors exploit the fact that a conductor through which current flows is deflected in a magnetic field by a force proportional to the current. This force vector is perpendicular to both the current vector and the magnetic field vector. $F = I(l \times B)$ applies, where I is the current vector, B the magnetic field vector and l the effective (= in the magnetic field) length of the conductor (see Fig. 9.5). Further information on electrical actuators can be found in e.g., [1].

9.3.1 DC Motors

Direct current (DC) motors essentially consist of the rotor (the rotating element) and the stator (the stationary element). The rotor includes the armature and change-over switch, while the stator comprises the permanent magnet(s) and the brushes (as an alternative to the permanent magnet, externally excited coils can also be used here). DC motors can be brushless or brushed. In brushed motors, the rotor is mechanically commutated via sliding contacts (brushes) to enable further rotation. In brushless motors, the external magnetic field is rotated under electronic control. Strictly speaking, these are AC motors which can, however, be controlled with the aid of their control electronics like DC motors.

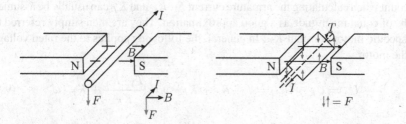

Fig. 9.5 Working principle of an electric motor

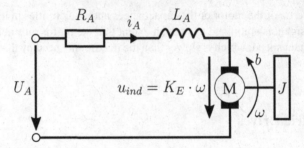

Fig. 9.6 Principle illustration of a DC motor

DC motors can be found, for example, in the window controls and seat adjusters of cars. They are also used in robotics for joint or platform drives. The advantages of these actuators are their simple integration into the mechanics, their relatively good controllability, the simple power supply and the very high actuating speeds.

The motor is operated by a current flow at the armature which creates a magnetic field in the opposite direction to the permanent magnet. The current flow in the rotor which lies in the magnetic field of the stator creates a torque at the rotor. The equivalent circuit diagram in Fig. 9.6 is used for the calculation. The torque T_A of the rotor remains constant and is calculated as follows.

$$T_A = K_T \cdot I_A \tag{9.3}$$

K_T is the torque constant of the motor and I_A the current through the armature. Rotation in the magnetic field induces a voltage U_{ind} at the armature that is proportional to the angular velocity ω. It is described in Eq. 9.4 using the magnetic flux Φ_F and specific motor constant c. Since the magnetic flux is assumed to be constant, the specific electric motor constant is simplified to $K_E = c \cdot \Phi_F$.

$$U_{ind} = c \cdot \Phi_F \cdot \omega \Rightarrow U_{ind} = K_E \cdot \omega \tag{9.4}$$

This voltage is called electromagnetic countervoltage. It acts as a damping of the motor. As it is proportional to the angular velocity, this damping also increases proportionally to the rotational speed. This damping must therefore be taken into account when calculating the armature current I_A. K_E and K_T can usually be assumed to be of equal magnitude as a good approximation. They are then simply referred to as specific motor constant K_F. In general, the following applies to the rated voltage at the motor

$$U_A = U_R + U_{ind} = I_A \cdot R_A + K_F \cdot \omega = \frac{T_A \cdot R_A}{K_F} + K_F \cdot \omega \tag{9.5}$$

Here R_A is the resistance of the armature and U_R the voltage applied to the armature. The difference between the two voltages U_A and U_{ind} corresponds to the voltage at the resistor R_A and the inductance L_A (see Eq. 9.6). For simplification, the magnetic field is assumed to be constant.

$$U_A - U_{ind} = R_A \cdot I_A + L_A \cdot \dot{I}_A \tag{9.6}$$

As can be seen clearly, the torque of the motor decreases with increasing rotational speed and is lowest in idling (idling current I_0) and highest in blocking ($T_{max} = K_F \cdot I_b$, with I_b as blocking current).

Let the mechanical power (output power) be $P_{mech} = T_A \cdot \omega$ and the electrical power (input power) $P_{el} = U \cdot I$. The maximum efficiency of the motor is given by $\eta = P_{mech}/P_{el}$. The rotational speed N of the motor is

$$N = \frac{\omega}{2\pi} = \left(\frac{U_A}{K_F} - \frac{T_A \cdot R_A}{K_F^2} \right) \cdot \frac{1}{2\pi} \tag{9.7}$$

This results in the mechanical power curve

$$P_{mech} = T_A \cdot \omega = \frac{R_A}{K_F^2} \cdot T_A^2 + \frac{U_A}{K_F} \cdot T_A \tag{9.8}$$

These relations are shown in Fig. 9.7. Since no mechanical work is performed at standstill and at maximum speed, the efficiency η is zero. However, at standstill the maximum torque T_b is applied, which is used in electric vehicles for a high initial acceleration. The torque decreases linearly with increasing speed. The relationship between current and torque is the opposite. At maximum torque T_b, i.e. at standstill, the largest current I_b, flows, since in this case the motor is virtually a short-circuit. It should be noted that the maximum mechanical power P_{max} generated is at a higher torque (approximately at $\frac{T_b}{2}$) than the highest efficiency η_{max} of the motor. This can be used to optimize efficiency or performance. The advantages of the DC motor lie

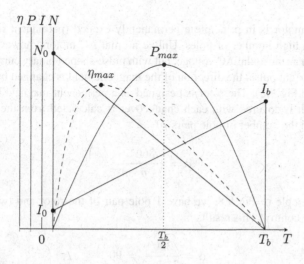

Fig. 9.7 Characteristic parameters of a DC motor

Table 9.1 Parameters of DC motors

Output power	$P_{mech} = T_A \cdot \omega$
Input power	$P_{el} = U_A \cdot I_A$
Efficiency	$\eta = \dfrac{P_{mech}}{P_{el}}$
No-load current	I_0
Blocking current	I_b
Idle speed	$N_0 = \frac{\omega_{max}}{2\pi} = \dfrac{U_A}{K_F} \cdot \frac{1}{2\pi}$
Maximum torque	$T_b = K_F \cdot I_b$
Rated voltage	$U_N = U_{A,N} = U_R + U_{ind}$

in the very good ratio between power and weight, the linear torque-speed curve and the higher peak torque compared to other motors.

In real operation, the counter torque of the load must be taken into account in addition to the torque generated by the motor. The torque of the movement of armature and load is described in Eq. 9.9. T_B is the difference between the drive torque and load torque and the moment of inertia of motor and load is described by J.

$$T_B = J \cdot \dot{\omega} \tag{9.9}$$

The most important formulas for motor calculation are summarized in Table 9.1.

9.3.2 Stepper Motors

The stepper motor is in principle a permanently excited (permanent magnet) AC motor with a high number of poles. Unlike normal AC motors, however, it is not supplied with a sinusoidal AC voltage, but with pulse-shaped binary current signals. With each current pulse, the direction of the magnetic field is changed by switching the coils (see Fig. 9.8). The rotor is designed as a permanent magnet. The angular step α, which is covered with each changeover is calculated from the number of phases m and the number of pole pairs p.

$$\alpha = \frac{360°}{2 \cdot p \cdot m} \tag{9.10}$$

In the example of Fig. 9.8, we have 1 pole pair of the rotor and two electrical phases of the control. This results in

$$\alpha = \frac{360°}{2 \cdot 1 \cdot 2} = 90°$$

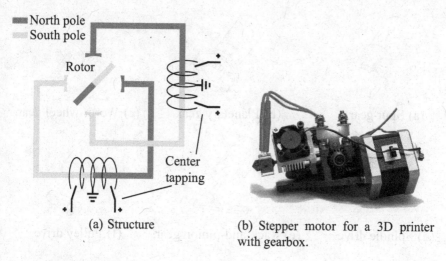

(a) Structure

(b) Stepper motor for a 3D printer with gearbox.

Fig. 9.8 Stepper motors

Low-pole stepper motors perform relatively large angular steps of $\alpha = 7.5°$ for example, while high-pole motors also allow step widths of $\alpha < 1°$. The achievable angular velocity ω depends on the frequency f with which the control pulses are sent to the motor.

$$\omega = \frac{2\pi f}{2 \cdot p \cdot m} \tag{9.11}$$

Step frequencies of up to 100 kHz are common. Stepper motors have the advantage that no further sensors are required to determine the angular position of the motor shaft and no controller for position and speed control. Since each pulse rotates the motor by a known angular range, only the necessary pulses have to be calculated. This applies as long as the external load torque does not exceed the tilting torque of the motor. If this is the case, the angular step is omitted and there are permanent deviations between target and actual positions. The forces generated are also comparatively low compared to other electric motors.

9.4 Gearboxes

Depending on the version, the speeds of motors range from a few hundred revolutions per minute to several tens of thousands, whereby the torque is usually relatively low at high rotational speeds. Gears are used to adapt the rotational and linear speeds and the torques and forces to the requirements of the application. Various gearboxes are known from the literature. Examples include planetary gears (Fig. 9.9b), screw or spindle drives (Fig. 9.9d) and harmonic drives (Fig. 9.11).

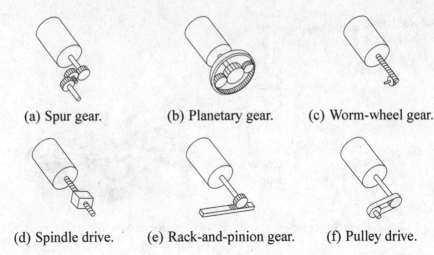

(a) Spur gear.   (b) Planetary gear.   (c) Worm-wheel gear.

(d) Spindle drive.   (e) Rack-and-pinion gear.   (f) Pulley drive.

Fig. 9.9 Various types of gears

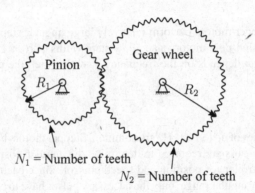

Fig. 9.10 Spur gear

Planetary gears are made up of several gear wheels of different sizes with the "planet"-gear wheels circling around a central "sun"-gear wheel, while an outer "ring"-gear holds everything together. In the case of spur gears, the drive shafts are parallel. In general, there is a linear relationship between two adjacent gears in a gearbox. In the following the number of teeth of the driving pinion is N_1 and the number of teeth of the driven gear wheel N_2 (as for example in Fig. 9.10 in a spur gear wheel), then the transmission ratio is $n = \frac{N_2}{N_1}$, also written as $N_2 : N_1$. The angular velocity of the driven spur gear is calculated as $\omega_2 = \frac{\omega_1}{n}$. The output torque is determined by $T_2 = n \cdot T_1$ (Fig. 9.11).

With chain drives, wire rope hoists and toothed belt drives, the transmission ratio, torque and angular speed can be calculated almost analogously. Unlike the planetary gear, the two pinions do not run directly against each other, but are connected by the above means of transmission. However, depending on the selected transmission, higher friction losses can occur here.

Fig. 9.11 Harmonic drive

With screw and spindle drives, the force and speed of movement generated depend on the pitch of the screw or spindle winding. With this type of drive, a rotary motion is converted into a linear motion. The pitch constant p, also called the pitch height, corresponds to the transmission ratio of spur gears. This is the distance that the screw travels during one revolution. If $v(t)$ is the linear velocity and ω the angular velocity, the result is

$$v(t) = p \cdot \omega(t)$$

Finally, Harmonic Drives are an excellent way to achieve a large gear ratio in the smallest of spaces. The operating principle of the harmonic drive is that an elliptical disc (wave generator) is located inside a flexible gear (flex spline), which in turn is located inside a fixed ring with internal toothing (circular spine). This flex spline has fewer teeth than the circular spline, which results in the transmission. Single gear stages can reach efficiencies of about 96%. However, to reach high transmission ratios in practice, several gear stages have to be stacked. As also other parts of the drive train like bearings and links are lossy, the overall efficiency is usually lower.

9.5 Other Actuators

A large number of alternative transducers have been developed to provide actuators for small actuating forces or torques in a light and space-saving way. Especially in the field of bionics, many biologically inspired technologies have emerged. However, due to their often low actuating forces, torques or deflections, they are only used for special applications.

9.5.1 Rheological Fluids

Rheological fluids, which occur both as *electrorheological* (*ERF*) and *magnetorheological* (*MRF*) fluids, can also be used to build actuators. The basic principle is that light oils with polarizable/ferromagnetic particles (20–50% of volume) change their flow properties depending on an applied electric/magnetic field. In general, the

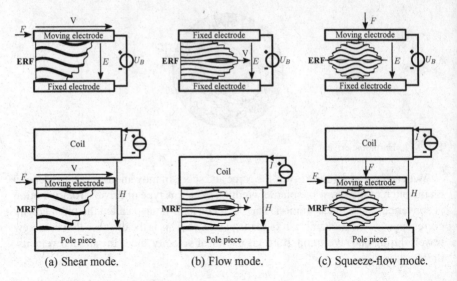

(a) Shear mode.   (b) Flow mode.   (c) Squeeze-flow mode.

Fig. 9.12 Basic modes of ERF/MRF

flow resistance increases with increasing electric/magnetic field strength, with reaction times of only a few milliseconds. When the field is switched off, the original properties return.

This can be exploited in different scenarios. In shear mode (Fig. 9.12a), two oppositely polarized electrodes move relative to each other. Depending on the electric field E, the free electrode moves faster or slower when a force F is applied.

In flow mode (Fig. 9.12b), both electrodes are fixed. They are used to influence the flow resistance of the fluid. The flow velocity is determined by the electric field E.

In squeeze-flow mode (Fig. 9.12c), the electrodes move towards each other. The squeezing flow builds up a pressure cushion between the electrodes, which is influenced by an electric field. This controls the force generated by the actuator. Two application examples of ERF are shown in Fig. 9.13.

The left figure (a) shows an ERF valve. Depending on the voltage applied by the high voltage supply, the flow velocity of the liquid in the cylinder is controlled. In principle, this simulates the opening of a valve. With the ERF shock absorber in Fig. 9.13b, which operates according to the shearing principle, the damping of the piston in the cylinder is adjusted by the control voltage. The higher the voltage, the more viscous the liquid and the greater the damping. Another example is the electrorheological disc clutch shown in Fig. 9.14.

It essentially consists of parallel plates (or concentric cylinders) located at the end of the shafts and surrounded by a rheological fluid. A (high) voltage is applied to the clutch discs via slip rings. By means of the field generated in this way, the viscosity of the ERF and thus the frictional connection between the clutch discs can be steplessly and very precisely controlled. This allows the transmitted torque to be

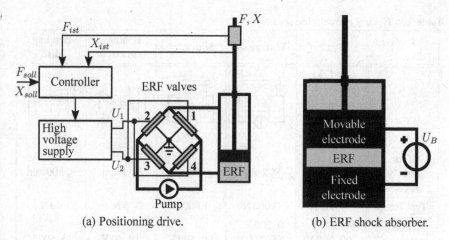

(a) Positioning drive.   (b) ERF shock absorber.

Fig. 9.13 Application examples of ERF valves

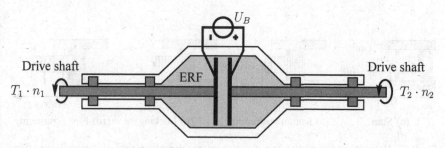

Fig. 9.14 Disc clutch based on ERF

set, as well as a transmission ratio. However, due to the basic viscosity of the ERF, this clutch cannot be fully opened, so that a minimum torque is always transmitted. This example can also be used for an application for rotating brake discs.

9.5.2 Piezoelectric Actuators

Piezo actuators (and piezo sensors) are based on the piezoelectric effect, which establishes the relationship between an electrical voltage that occurs and the deformation of certain solids. For the piezo actuator this means, that a deformation of a crystal is produced by application of a high voltage. This means that electrical energy can be quickly converted into mechanical energy without the need for moving parts. Piezoelectric actuators are characterized by very fast response times and a long service life. Due to the very low leakage currents, the electric field is maintained without energy supply, which means that an energetically favorable actuator can be constructed. The disadvantages are the very low deflections of the actuator, but these can be adjusted

Table 9.2 Types of piezoelectric actuators

	Stack	With gears	Slip-Stick	Bending transducer	Bending disc
	$\updownarrow\Delta x$	$\Delta x \updownarrow$	$\updownarrow\Delta x$	$\Delta x \updownarrow$	Δx
	U_B	U_B	U_B		
Typ. travels/ strokes	10..200 μm	$\leq$ 2 mm	$\leq$ 50 mm	$\leq$ 1 mm	$\leq$ 500 mm
Typ. force	30.000 N	3.000 N	1.000 N	5 N	40 N
Typ. operating voltage	60..200 V 200..500 V 500..1000 V	0..200 V 200..500 V 500..1000 V	60..500 V	10..40 V	10..500 V

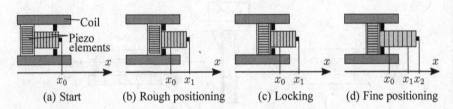

(a) Start   (b) Rough positioning   (c) Locking   (d) Fine positioning

Fig. 9.15 Hybrid transducer consisting of a solenoid (rough positioning) and a piezo element (Fine positioning)

very precisely. By means of appropriate combination of single piezo elements a variety of actuators can be created depending on the application (see Table 9.2). They are also often used as part of a complex actuator system, such as the linear positioning drives shown in Fig. 9.15.

In this case, an electromagnet is first used to roughly position the whole work carriage, which is made up of piezo elements (see Fig. 9.15(b)). This position is then fixed by actuating the piezo elements mounted perpendicular to the direction of movement. They expand and clamp the carriage, as shown in Fig. 9.15(c). Finally, the piezo elements fitted in the direction of motion are excited to such an extent that the fine position required is achieved (see Fig. 9.15(d)).

Magnetostrictive materials are used in a similar process, in which length changes depend on the change in the magnetic field. The expansions lie in the per mill range (1.2 mm/m). These elements have a high actuating speed and very short reaction

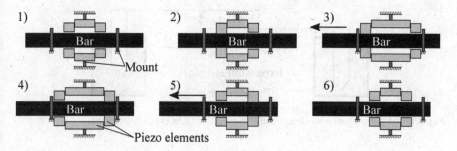

Fig. 9.16 Inchworm motor

Table 9.3 Applications of piezoelectric actuators

Optics	Medicine/Biology	Precision mechanics	Mechanical engineering
Laser calibration systems	Braille readers	Micro manipulator (Tunnelling microscope)	Positioning in lathes/milling machines
Positioning of optical fibres	Metering devices, Micro nozzles	Linear-/rotation motors	Non-round turning, drilling, grinding
Adaptive optics (e.g., Mirror-arrays)	Creation of shock waves	Video head tracking system	Fast control of brakes, locking devices
Mirror deviation, ray tracking	Lithotripter	Ink-jet printers	Readjustment of tools for the compensation of abrasion
X-ray-/micro- lithography	Ultrasonic-scanner	Speaker (e.g., telephone)	Active oscillation damping
Autofocus systems	Ultrasonic cutting or cleaning devices	Injection valves (motor vehicle)	

times (ms range). About three-quarters of the magnetic energy is converted into mechanical energy. However, very high losses often occur in the power amplifiers.

Compared to piezo crystals, the effect can be exploited at higher temperatures (up to approx. 4000 °C), the elongation hysteresis is lower and there is no moving electrode. However, ohmic losses due to magnetizing currents also occur in the static case. An additional technical problem is that the integration of the magnetic field is difficult due to low permeability of commonly used materials, such as Terfenol-D.

A possible actuator with piezoelectric components is the *Inchworm motor* shown in Fig. 9.16. A smooth shaft is axially displaced over 6 phases. By holding one end, contracting or expanding the piezo element and then gripping around it, the shaft is moved on step by step.

Piezo actuators are now used in various applications in medicine, optics, precision mechanics and mechanical engineering. Examples are injection valves in the automotive sector and elements for active vibration damping. Further examples are listed in Table 9.3.

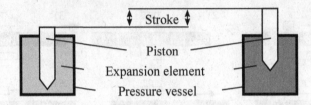

Fig. 9.17   Moving a piston by changing the temperature of an expansion element

9.5.3   Thermal Actuators

Another way to build actuators is to use heat as primary energy. For this purpose, for example, a heating coil is controlled by the control signal of the processing unit, which in turn causes a change of state in the actuator.

9.5.3.1   Expansion Elements

Expansion elements belong to this class of thermal actuators. They exploit the fact that the volume of solids increases with increasing temperature. The typical design is as follows. In a piston there is a pressure vessel filled with expansion material (e.g., wax), as can be seen in Fig. 9.17. When the wax element is heated, there is an increase in volume which causes the piston to move upwards. A return spring allows the piston to be returned to its initial position when the system cools down. Expansion elements can have a linear temperature-stroke characteristic (small and large control temperature ranges possible, e.g., 150°C/1500°C) and a non-linear one. With non-linear characteristic curves, the stroke moves in jumps within a small temperature range (e.g., 1°C … 2°C). Application examples for expansion elements can be found in the automotive sector, such as cooling and oil circuit control.

9.5.3.2   Bimetals

Bimetals extend the idea behind the expansion elements. Here, two metal strips with different coefficients of thermal expansion are joined together. When this bimetal is heated, one side expands more than the other, resulting in an overall bending of the bimetal strip. As a result, bimetals are suitable as temperature indicators or overheating switches, for example, but can only perform comparatively little work per volume unit. In addition, the structure as a layered composite material only allows bending deformations and also sets narrow limits to the shaping. The deformation occurs continuously with the temperature change.

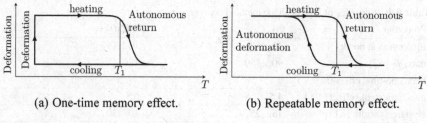

(a) One-time memory effect.   (b) Repeatable memory effect.

Fig. 9.18 Deformation of shape memory alloys

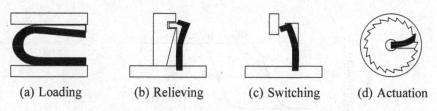

(a) Loading   (b) Relieving   (c) Switching   (d) Actuation

Fig. 9.19 Applications of shape memory alloys—by heating or cooling, the shape changes

9.5.3.3 Shape Memory Alloys

Shape memory alloys change shape depending on a fixed deformation temperature. A distinction is made between shape memory alloys with a one-time memory effect and those with a repeatable memory effect (see Fig. 9.18).

With the one-time memory effect, the component is permanently deformed at low temperature. When heated above a threshold temperature, the component returns to its original shape. Typically, the temperature range for forming is between 100°C and 200°C. During conversion, the material performs work and can thus be used as an actuator (see Fig. 9.19).

By thermomechanical pre-treatment of the material, different shapes can be achieved in the cold and hot condition. This is called the repeatable memory effect. The work can only be done during heating. Shape memory alloys are metal alloys such as nitinol (NiTi) or copper-based alloys (e.g., CuZnTi) (see Table 9.4). Polymer compounds with similar properties are now also known. These have a shorter switching time, but can only be used to a limited extent as actuators due to lower restoring forces.

Shape memory alloys must not be confused with bimetals. The latter do not have the pronounced hysteresis of shape memory alloys (see Fig. 9.20), but change their shape continuously with temperature. In addition, bimetals can do significantly less work per volume used and are more limited in shape and degrees of freedom. But they are much cheaper. Thus bimetals are better suited for tasks requiring continuous deformation with low force, while shape memory alloys are advantageous in applications where it is necessary to switch between two dedicated states. Further information on the use of shape memory alloys can be found e.g., in [2].

Table 9.4 Properties of typical shape memory alloys

Properties	NiTi	CuZnTi
Elongation at break [%]	40...50	10...15
max. T_1 [°C]	90...150	120
max. one-time effect [%]	6	4...5
max. repeatable effect* [%]	4.5	0.5...1
Hysteresis width [K]	15...25	10...20
Long-term stability	Good	Worse

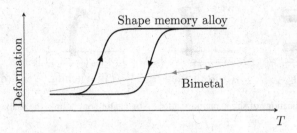

Fig. 9.20 Difference of shape memory alloys and bimetals

References

1. Lee EA, Seshia SA (2016) Introduction to embedded systems: a cyber-physical systems approach. MIT Press, Cambridge
2. Lagoudas DC (ed) (2008) Shape memory alloys: modeling and engineering applications. Springer Science+Business Media, LLC, New York

Chapter 10
Processing Units for Embedded Systems

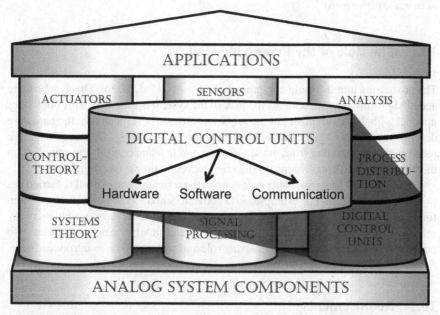

This chapter introduces different architectures for the computing/processing unit(s) of embedded systems. Based on requirements such as real-time capacity, robustness and energy consumption, different available units are compared. The microcontroller as one of the most commonly used processing units and the FPGA as the main representative of programmable hardware are considered in more detail and their integration into the environment is explained using a sample circuit.

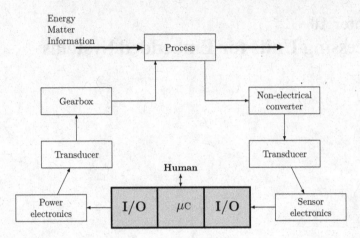

Fig. 10.1 The processing unit in the measurement and control cycle (here, a microcontroller with its input and output ports)

10.1   The Role of the Processing Unit

The processing unit is a main component of the embedded system. This is where the actual algorithm that implements the task of the system is executed. This book does not deal with the development of the required software or respectively, the hardware description. For this we recommend for example [1, 2] and [3]. The processing unit receives the data coming from the sensors, interprets it and controls the actuators on the basis of logical conclusions (in different representations) (see Fig. 10.1).

Often a part of the sensor data processing or the actuator control itself is carried out in the processing unit. For example, A/D converters and PWM generators are often found in microcontrollers. Software-programmable systems have a high everyday relevance for most users due to the spread of PCs and smartphones, but there are also other architectures that play an increasing role and will therefore be introduced here.

10.2   Requirements

The architecture chosen for an embedded system depends largely on the requirements it has to meet. In particular, the following non-functional properties play an important role.

- **Real-time requirements**   Embedded systems often have high requirements for execution speed and real-time capability, especially in safety-critical systems. Communication between individual nodes must also usually have real-time capability and be robust. Specialized and hard-wired architectures have an advantage over generic ones.

- **Weight, size, energy consumption** These requirements play a major role especially in mobile or autonomous systems. Highly integrated processing units are therefore often used.
- **Reliability, maintainability, availability** Since embedded systems are often used in industrial environments and are not designed for the direct interaction of the processing unit with humans, the requirements of reliability and availability are very high. Compared to consumer products, for example, it is not an option to restart a power plant several times a day because there is a variable overflow somewhere.
- **Robustness and EMC** *Robustness* should also be mentioned in this context, as embedded systems are often used in environments exposed to dirt, liquids or vibration which they must be able to withstand. A special requirement here is *electromagnetic compatibility* (*EMC*). On the one hand, this refers to the system's insensitivity to interference from external electromagnetic fields, but on the other hand it also relates to the fact that it does not unintentionally generate any such interference itself.

Of course, the requirements mentioned are weighted differently depending on the system. For example, a wireless motion detector does not require a powerful CPU, but has high requirements in terms of energy efficiency and space utilization. A computer network for weather simulation, on the other hand, has a sufficient energy supply and enough space, but has high demands on speed and availability.

10.3 Processing Units—Overview of Types

Which algorithm is actually executed plays no role in the hardware design, but it does play a role in selection of the processing unit. A wide variety of types can be found here, such as PCs, industrial PCs, programmable logic controllers (PLCs), microcontrollers/-processors, digital signal processors (DSPs), field-programmable gate arrays (FPGAs), application-specific integrated circuits (ASICs), application-specific instruction-set processors (ASIPs) and many more. These can be classified according to their configurability (are the systems hard-wired, partially wired or self-configurable?), availability (is the system available as a finished off-the-shelf component or is it only the basis for further development?) or application proximity (is the system as generic as possible, or tailored to certain applications?), as shown in Fig. 10.2.

Furthermore, there is a wide range with regard to the power of the systems. For example, a processing unit for a motion detector can consist of an IC with 4 legs only, while powerful ARM processors can be used for modern smartphones. Some common technologies will be presented below with examples of their application.

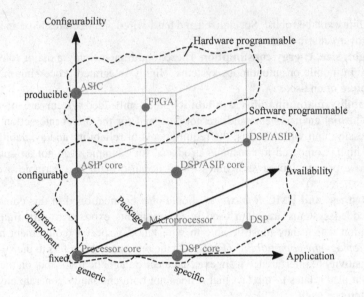

Fig. 10.2 Classification of different hardware technologies

10.3.1 Generic, Software-Programmable Processing Units

Most embedded systems today use software-based processing units such as PCs, industrial PCs or microcontrollers. These are general-purpose building blocks that can be programmed by software for a variety of different applications. They have a hard-wired architecture based on fixed processor cores. Due to their generic nature, they are easily interchangeable and very inexpensive, but in many cases they do not deliver optimal performance because certain resources are not used or are not available. The PC is a good basis for development. As a relatively powerful computer, it can be used to control stationary or developing systems reliably. Real-time capability can also be guaranteed by appropriate operating systems.

However, the classic PC is quite sensitive to dirt and vibration and is a rather bulky component. Despite its internal modularity, the architecture also often results in unused resources that require unnecessary space or energy. Most motherboards have more bus connections than required and many applications do not require a sound or graphics card.

One way to avoid this is to use an *industrial PC(I-PC)*. It has a more modular structure than a standard PC, so that some of the otherwise permanently installed peripherals can be replaced and adapted to the application. In addition, its design provides special protection against vibration and dirt, making it ideal for use in industrial environments. Many production plants are controlled by industrial PCs, but they are also used in commercial vehicles.

Figure 10.3b, c show different designs for industrial PCs. The PC/104 module has been an IEEE standard since 1992 and exists in several hundred versions. The goal

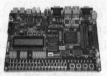

(a) Digital signal processor.   (b) Industrial PC.   (c) Industrial PC: PC/104.   (d) FPGA development board.

Fig. 10.3 Different processing units: **a** DSP (MC56F8357), **b** Industrial PC in a case, **c** Industrial PC stack (PC/104), **d** FPGA development board (Intel/Altera Cyclone IV)

of this standard is to create a compact, standardized format for modular, PC-based systems, especially for industrial applications.

PC systems are not suitable for applications that have special requirements in terms of space, energy consumption or weight, as they are quite extensive due to their high genericity and focus on interaction with people. Microcontrollers currently play an important role here. These are also general-purpose units, but in addition to the processor they also combine important peripheral elements (e.g. bus drivers, PWM units, D/A converters, often also RAM and ROM etc.) in one chip.

This offers the advantage that the peripheral elements have the optimal connection to the processor. The whole system becomes very small and light and power consumption is reduced. Due to the wide range of available microcontrollers, the hardware can be optimized to a large extent by the selection.

In Fig. 10.7, a microcontroller of the ATmega family can be seen. In general, microcontroller-controlled systems have a lower computing power than PC systems (whereby the transitions become increasingly blurred) and often a limited instruction set. This will be discussed in more detail in Sect. 10.4. Microcontrollers are used not only in smartphones and microcomputers, but also in sensor networks, vehicle controls, kitchen appliances and much more. More detailed information on the design and application of processors and microcontrollers can be found e.g. in [3, 4] and [2].

10.3.2 Application-Optimized, Software-Programmable Processing Units

In many cases certain algorithms or algorithmic structures are used repeatedly. Here it makes sense to also adapt the hardware of the executing processing unit accordingly and to optimize it for these structures. This allows a higher efficiency and/or performance compared to generic computers.

Digital signal processors (*DSPs*) are basically microcontrollers with special peripherals and an extended instruction set. They are optimized for signal processing tasks such as digital filters or other signal manipulations. They are therefore frequently used in audio and video applications, but also serve as accelerators for communication in distributed systems. For this purpose, they usually contain built-in D/A and A/D converters that are directly connected to the processor. An essential element of DSPs is the presence of several processing units, including a *multiply-accumulate unit (MAC)*. This allows simultaneous execution of addition and multiplication as required for convolution and Fast Fourier Transformations (FFTs) (see Sects. 5.3 and 5.4.1), which play an important role in signal transmission and coding in particular. Figure 10.3a shows a DSP (MC56F8357) of this sort. Further information on the design and development of DSPs can be found in [5].

Another special component is the *programmable logic controller (PLC)*. These processing units are specially designed for use in cyclical production plants. As a rule, they are real-time capable systems that can be built up in modules. They can take the form of modules on a DIN rail, for example, or software on an industrial PC.

PLCs are programmed with special languages (e.g. ladder diagram, function blocks, IL, etc.), which are usually designed for logical circuits or automatic machines. These programs are processed cyclically in the PLC according to fixed patterns, which guarantees real-time capability on the one hand and reduces the development effort on the other. A comprehensive overview of the application and special features of PLCs can be found in [6].

10.3.3   Application-Specific, Hardware-Programmable Processing Units

If such prefabricated, programmable devices are not sufficient to meet the requirements for energy efficiency or general performance, the use of *Application-Specific Integrated Circuits (ASICs)* may be a suitable solution. These are chips specially manufactured for a task in which the algorithm is integrated into the hardware. They are optimized with regard to the task-specific requirements for speed, energy efficiency, size and/or reliability, but are usually also very inflexible.

The main problem is that development and production in small quantities are very expensive and time-consuming, which is why ASICs are only cost-effective if mass production is planned. Many ASICs can therefore be found as freely available standard parts from the respective dealers; these are referred to as *Application-Specific Standard Products (ASSP)*.

For example, A/D converters, USB interfaces, step-down voltage sources and display drivers are available as individual ICs. The specific application is still relevant and not, for example, the size or shape of the components (see Fig. 10.4). For example, the NE555, which is a monostable, unstable or bistable flip-flop depending on the wiring and can also be used for other applications, would not be referred to as an ASSP.

Fig. 10.4 Different IC packages

10.3.4 Generic, Hardware-Programmable Processing Units

Often the problem arises that the general requirements, such as parallel data processing, time optimization or real-time capability, would actually indicate using an ASIC, but the development of such an ASIC would not be economical in small quantities. In these cases *Field-programmable gate arrays* (*FPGAs*) are used. This area of configurable hardware represents one of the fastest growing market segments for embedded systems.

These are special ICs which have not yet learned any concrete behavior. This means that you can "program" the hardware itself. In contrast to a microcontroller, you do not specify the behavior of established hardware by means of software, but change the hardware itself. This offers many options for using parallel signal processing or custom instruction sets. In contrast to ASICs, FPGAs can be reconfigured several times, which is an advantage not to be underestimated, especially during development. FPGAs are also used for the development of ASICs, since the temporal interaction of the ASIC components can be simulated on them. Compared to microcontrollers, FPGAs are quite expensive and perform slightly less well in terms

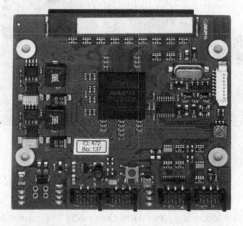

Fig. 10.5 CPLD-Board (Intel/Altera MAX II)

of cycle time with the same manufacturing technology. The development effort is also greater. More details can be found below in Sect. 10.5. Figure 10.3d shows a development board for an Intel/Altera Cyclone IV-FPGA, on which the required peripherals and various input and output options are already integrated. *Complex programmable logic devices* (*CPLDs*) are a cheaper alternative. One of these is shown in Fig. 10.5. They are similar in structure to FPGAs, but usually have a significantly more limited range of functions. For a more detailed introduction to working with FPGAs, we recommend e.g. [6] and [5].

10.4   Microcontrollers

This section explains the basic structure and function of software-programmable processing units using the microcontroller as an example.

Software-programmable systems still make up the majority of products. With the introduction of smartphones, individual families such as ARM processors have become widely known. Figure 10.6 shows a processing unit of this sort (BCM2837RIFBG with ARM Cortex A53 cores) on a Raspberry Pi. On this single-board computer, it performs the tasks of a CPU (and partially GPU). The Raspberry Pi is widely used in prototype development of embedded systems because it provides not only the components required for pure operation of the processor, but also extensive peripheral elements such as graphic decoders, A/D converters, USB drivers, WLAN and directly usable I/O ports. Software-programmable systems have been in use for decades through the use of PCs, so that there is a broad base of user knowledge and a large developer community.

Fig. 10.6   Raspberry Pi 3 Model B with an ARM Cortex A53-based controller

Fig. 10.7 ATMEGA32-Microcontroller on an Arduino Nano-Board

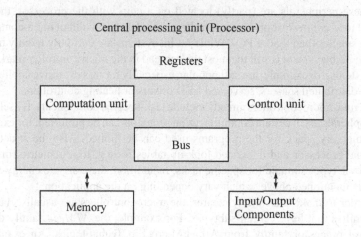

Fig. 10.8 Structure of a processor

The basic idea of this control technology is to create hardware that can execute as wide a range of algorithms as possible in an acceptable time. For this purpose, a processor is used which processes the programs and data provided in the software.

In principle, a processor consists of a *control unit* which decodes the program instructions and converts them into corresponding calculations, and an *arithmetic unit* which then executes them. Registers store intermediate data and a bus connection enables communication with the rest of the system (memory, input/output, other peripherals) (see Fig. 10.8).

Since embedded systems are usually created for a specific function, such general-purpose processors are extended to include application-specific peripherals (memory, communication interfaces, PWM generators, etc.) (see Fig. 10.9).

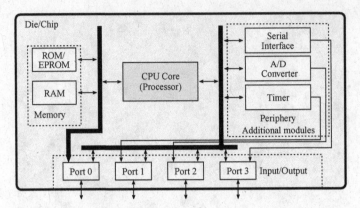

Fig. 10.9 Example of the structure of a microcontroller, all peripherals are on the same silicon die as the processor

If these peripherals are (partly) located on a chip with the processor, they are referred to as *controllers* or *microcontrollers*. The data bus width of microcontrollers is often smaller than that of PCs. While 64-bit systems are currently mainly used in PCs, the 32-bit system is still the most widespread in the microcontroller market and 8-bit systems, for example, are still popular, especially for less extensive applications such as distributed sensors, servo and LED drivers or heating controllers.

As a rule, microcontrollers already include task-specific components. By selecting the peripherals used, certain communication protocols can be preferred, for example, or the memory space for the programs used can be limited. Also the structure of the actual processor and integrated look-up tables for e.g. trigonometric functions make some types suitable for specific tasks. In addition, there are the energy-saving concepts mentioned above, which vary depending on the application.

In order to make this variety clearer, the microcontrollers are usually classified into families that have certain subtypes. For example, the *ATMega* family defines a general processor family from Atmel/Microchip Technologies. An example is depicted in Fig. 10.7. From this you can choose between different memory sizes (ATMega8/16/32/64/128 each in KByte), and also different peripherals and energy saving modes (e.g. ATmega168P: 8-bit AVR microcontroller, 16 KB Flash, 28/32 pins, with special power saving functions). Other options would be different voltage levels or maximum clock rates.

A standard *microcontroller chip* contains serial/parallel interface(s), fieldbus interface(s) (see Sect. 11.6.2), program and data memory, and A/D and D/A converters (see Sect. 7). The most compact microcontrollers do not even require an *external clock generator*, but can generate a clock internally for simple tasks. On the other hand, due to the integrated peripherals, they are also much more specialized in their possible application. They are therefore usually *expandable* by external circuitry to attach additional memory, additional I/Os, etc., for which a system bus is used that leads to the outside (i.e. to the pins of the microcontroller). They also have internal

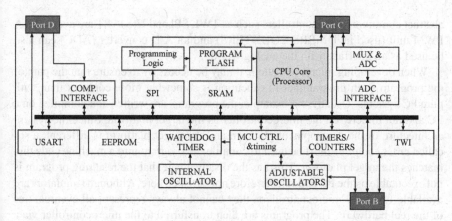

Fig. 10.10 Principle structure of an ATmega8 by the company Microchip Technologies Inc

program memory. In mass production, a *controller operating system* is frequently integrated as ROM. Often you will also find versions with integrated *(E)EPROM* or *Flash* for application programs, as otherwise the program memory would have to be connected externally. The programming of microcontrollers differs from the programming of PCs in some respects. The limited hardware of the microcontroller must be taken into account, which is particularly reflected in the addressing range and the clock rate. Due to the smaller community for individual controllers, the choice of usable software libraries is also smaller.

Because of the often lower performance and the fact that direct interaction of humans with the hardware is required, hardware-related programming languages play a more prominent role in the area of embedded systems. In addition to the most common programming languages, such as C/C++, in many cases (sub-)programs are also written in Assembler. One aspect that is also reflected in the programming is the frequent use of interrupts. These special events are usually triggered by external factors and cause the microcontroller to interrupt the current program execution, to push all relevant data into the memory and to jump to the *interrupt routine*. This is used to react to the corresponding event. It then jumps back to the actual program. This is made possible by special interrupt controllers, which form an additional peripheral element in the microcontroller. Although this allows fast reactions to external influences, problems arise with the real-time capability of the embedded system, which is explained in more detail in Sect. 13.

An example of such a microcontroller is the ATmega8 shown in Fig. 10.10. The grey box shows the actual CPU. The chip also contains fixed program memory (Program Flash or EEPROM) and internal RAM (SRAM). An internal oscillator (ADJUSTABLE OSCILLATOR) is available as a clock generator, but more precise external clock generators can also be connected. The watch dog timer (WDT) should also be mentioned in this context, which triggers a reset if the system crashes. Various protocols are available for communication with the outside world. On the one hand the external pins of the microcontroller can be used as simple I/O ports, on the other hand abstract communication controllers such as TWI, SPI and U(S)ART are available. A PWM unit (used in TIMERS/COUNTERS) and an A/D converter (ADC) can also be used as an interface with the analog world.

When developing microcontrollers, it may be necessary to ensure that the path of the program from programmer to execution is somewhat more complex than with pure PC systems. Usually, for example, the programs are written and compiled on a PC and not directly on the microcontroller as the target platform. Since the system architecture of the development platform and the target platform is different, a so-called *cross compiler* must be used to compile the program into a machine code that matches the target platform. This has the disadvantage that the resulting program is not executable on the PC and can therefore not be tested here. Although simulators are available for many microcontrollers, they cannot always represent all eventualities of the real hardware. The programs are then transferred to the microcontroller via a programming device with a corresponding interface (JTAG or similar). For testing on the real system, a complex debugging toolchain must be used if the PC is to continue to serve the visualization and stimulation of the system. This toolchain not only enables step-by-step execution of the microcontroller program, but also allows access to its hardware resources and provides a communication interface with the PC. This creates additional sources of error, which makes troubleshooting more difficult.

Irrespective of the choice of the actual processing unit, it must also be integrated into the environment in the end. Therefore the wiring of a microcontroller with a push-button (input) and an LED (output) is described here as an example, see Fig. 10.11.

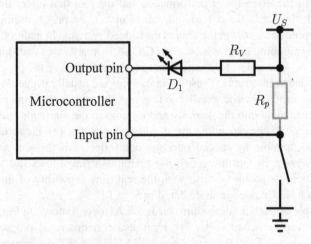

Fig. 10.11 Wiring of the LED and the push-button

The LED should light up when the corresponding output pin is switched to '0', i.e. the low voltage level (usually GND) is connected. The push-button should draw the signal at the input pin which is actually on U_S ('1') to GND ('0'). This requires some information about the components used. In this example, it is assumed that the external circuitry and the microcontroller itself operate at the same voltage level of 5 V and are connected to the same ground. The output pins must be able to *drive* the current (i.e. not only physically to withstand the expected current strength, but also to maintain the desired voltage) that flows through the LED or push-button.

As we can see in Fig. 10.11, one pin was defined as the output and one as the input on the microcontroller. This must usually be done explicitly via software. Internally, most freely usable pins (often referred to as *General-Purpose-I/O-Pins*) are connected in such a way that they can assume three (simplified) voltage states: High, floating and low. While high and low correspond to the voltages U_S and GND, the floating state is located approximately between the two, but can be easily influenced from the outside, i.e. it adapts to the external circuitry. This can be helpful to follow values from external elements quickly, but also increases the susceptibility to interference from electrical fields in the environment.

In this case this is especially important for the input pin. When the switch is closed, its level should be pulled to GND, which means that when the switch is open, it must set a high voltage level itself. This can either be solved by an internal controller circuit, or, as shown in gray in the figure, explicitly by an external "pull-up" resistor R_P. If the switch is open, no current flows, which is why no voltage drops across the resistor. Thus the voltage level of the pin is equal to that of U_S. For this to work, the input resistance of the input pin must be large enough for any flowing currents to be negligible here.

When the switch is closed, a circuit is made between U_S, the pin and GND, as a result of which the voltage drops completely across the pull-up resistor and the voltage of the pin is GND. This resistance must be quite large to prevent too large a current flowing (see Ohm's law, Sect. 2.6). Usual values are around 10 kΩ. The current through the input pin in this circuit must be limited by the microcontroller itself.

In order to connect the output pin correctly, it is necessary to check whether the microcontroller can drive the current flowing through the output pin. For this an LED (D_1) with a forward voltage of 1.8 V and a forward current of 20 mA is assumed. Since a serial circuit is present in this branch, this is already the current to be driven. Now the series resistor R_V has to be designed appropriately so that the voltage level is correct. According to Ohm's law, for a value of R_V the following applies.

$$R = \frac{U}{I} \rightarrow R_V = \frac{U_S - U_{D1}}{I_{D1}} = \frac{5\,\text{V} - 1.8\,\text{V}}{20\,\text{mA}} = 160\,\Omega$$

If necessary, the internal resistance of the pin must also be taken into account. In this example, however, it plays a subordinate role.

Not so the capacitance of the input pins. It is tempting to work on these inputs with appropriately dimensioned voltage dividers and very high resistances. In this way you can cleanly transfer a voltage level (which is a logical '0' or '1') between digital components without loading the pins with a large current.

The problem is that each input has a capacitance. This is partly intended, for example, to smooth the input signals, but partly it results from parasitic effects. In the case of a changeover (between '0' and '1') an RC element is formed from the upstream resistor and the input capacitance. As shown in Sect. 3.5, it takes some time for such RC elements to be charged, i.e. until the desired voltage is applied.

Since communication between components requires the fastest possible signal transmission—i.e. many signal edges coming in quick succession—a large resistance is disadvantageous here. In practice, it must therefore be adapted to the specific case in such a way that it is large enough for the current is not to be too high, but small enough for it not to interfere with signal transmission.

10.5 Field-Programmable Gate Arrays (FPGAs)

As mentioned above, the number of processing units based on configurable hardware is increasing rapidly. Unlike microcontrollers, this is no fixed hardware on which various software programs can be executed, but the internal wiring and the function of the hardware itself is changed so that it executes the algorithm.

The most important representatives of this type are *FPGAs* (*field-programmable gate arrays*). As the name suggests, they consist of a two-dimensional array of logic blocks (LB) that can be connected by a programmable network of communication lines. Internally, the logic blocks consist of look-up tables (LUT) for the implementation of switching networks and flip-flops, which can be programmed and connected depending on the function. The concrete interconnections and internal implementations differ depending on the technology and manufacturer. In addition, special function blocks such as multipliers and memory modules are often already integrated. The basic structure is shown in Fig. 10.12.

The connections within the LBs and between the LBs, as well as the LUTs, can be set via a hardware description or a program, whereby a network of hardware elements is formed which fulfils the function required. The multiplier and storage elements can also be integrated to save space and increase efficiency.

This enables an almost unlimited number of applications. FPGAs can be used to execute individual algorithms fully in parallel, but programmable multiprocessor systems with self-defined instruction sets can also be implemented on FPGAs.

In this context, so-called *IP cores* (*Intellectual Property*) are playing an increasingly important role. These are individual program modules created by third parties that can be integrated into your own design, as would be the case for normal hardware design with purchased ICs.

These IP cores can contain highly optimized algorithms for certain tasks (especially communication), or implement proprietary protocols that should not be disclosed. Precreated *soft cores*, i.e. partially standardized microprocessors/controllers that are instantiated on the FPGA, are often used.

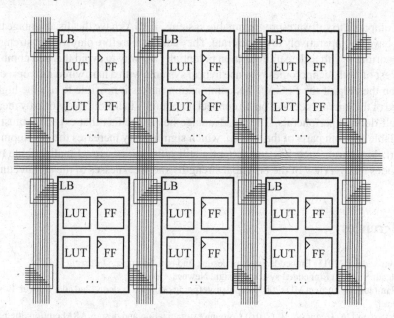

Fig. 10.12 Principle structure of an FPGA

Table 10.1 Comparison of the presented processing units according to the requirements for embedded systems

	Real-time requirements	Weight, size, energy consumption	Reliability, maintainability, availability	Robustness, EMC
PC	−	−	+	−
I-PC	−	0	+	++
µC	0	++	0	+
DSP	+	+	+	+
PLC	+	0	+	++
ASIC	++	++	+	++
FPGA	++	+	+	+

In addition to the development of ASICs, FPGAs are now also used in many areas that were previously reserved for software-programmable systems. They are playing an increasing role in real-time data processing of, for example, communication protocols, data encryption and multimedia data processing. They are also becoming increasingly important for the control of time-critical industrial plants, as they have the advantage over other systems of being reconfigurable, so that technical innovations can be reacted to with a reconfiguration of the FPGA without having to replace modules.

Compared to software-programmable systems, FPGAs have the disadvantage that they have a comparatively low clock rate. They cannot therefore play to their strengths in algorithms with little parallelization potential. One way to avoid this and combine the best of both technologies is to use hybrid systems with a hardwired microprocessor on the chip of the FPGA. This combination makes it possible to use the higher speed of the microprocessor, but to connect it optimally to the outside world by means of self-defined peripherals. The disadvantage lies in the fact that expertise must be available for both parts of the system, which significantly increases the development effort. Moreover, such ICs are currently still quite expensive. Finally, Table 10.1 provides an overview of the principal strengths and weaknesses of the architectures.

References

1. Marwedel P (2011) Embedded system design, 2nd edn. Springer, Dordrecht [i.a.]
2. Heath S (2002) Embedded systems design. Newnes, Oxford
3. Hamblen JO, Furman MD (2001) Rapid prototyping of digital systems, 2nd edn. Kluwer, Boston [i.a.]
4. Patterson DA, Hennessy JL (2016) Computer organization and design ARM edition: the hardware software interface. Morgan Kaufmann, Amsterdam [w.o.]
5. Liu D (2008) Embedded DSP processor design, vol 2. Systems on silicon. Morgan Kaufmann, Burlington
6. Woods R, McAllister J, Lightbody G, Yi Y (2008) FPGA-based implementation of signal processing systems. Wiley, Chichester

Chapter 11
Communication

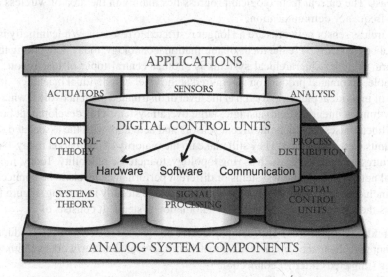

This chapter introduces the special requirements for communication systems in embedded systems and their solutions. The OSI layer model is presented and its lower two layers, the bit transmission layer and the link layer, are examined in more detail. In this way, the connection between the underlying electrical components and the abstract messages transmitted via them is conveyed. The correct selection and design of the transmission medium allows messages to be transmitted between stations largely without interference, and errors that occur can be detected and corrected by means of failsafe procedures. In addition, media access is addressed so that the communication of several bus participants can be controlled. The bus protocols Profibus and CAN are introduced as examples of such communication systems.

© The Author(s), under exclusive license to Springer Nature Switzerland AG 2021
K. Berns et al., *Technical Foundations of Embedded Systems*, Lecture Notes in Electrical
Engineering 732, https://doi.org/10.1007/978-3-030-65157-2_11

11.1   Communication Components of Embedded Systems

Embedded systems consist of a multitude of individual components, such as sensors, actuators and processing units. Furthermore, they are often integrated into a larger context. A motor control is part of an industrial robot that is part of a production line located in a factory that ultimately belongs to a company. At and between all these levels, the components must interact with each other to ensure a meaningful process. A key aspect here is the communication infrastructure, which makes interaction possible in the first place.

The constant increase in automation in all technical areas has also led to an almost explosive increase in demand, but also in possibilities on the communication side. More and more bus systems—some proprietary, some openly standardized—have been developed. Today there are already over 100 different communication systems in automation technology. Their number is still increasing, although not as fast as in the past. The current technological progress lies mainly in the area of wireless and high-frequency communication.

Wireless sensor networks are no longer restricted to research. While initially individual bus systems were the focus of automation technology, today networking them to form complex, hierarchical systems is another central topic of discussion. For example, automated production is required at the level of individual machines, at the level of individual plants, and even at the level of distributed plant networks, whereby everything is integrated into a single, large overall system. This development began with linear buses to which intelligent sensors and actuators with the associated control units were connected. They still make up the majority of the market today. Other structures followed later, such as ring topology, to increase reliability. Today, hierarchical networks also allow systems of different performance levels to be connected. The inclusion of the Internet in these networks is also steadily increasing. Figure 11.1 shows the abstract structure of digital automation systems. It consists of

- **Backbone (open Bus)** This comprehensive bus is used to network the production plant with the management level or to network several factories/works via Ethernet or a (tunneled) Internet connection.
- **Cell bus (remote bus)** This bus is used on the one hand to connect the cells to each other and on the other hand to connect the display and control elements with the process-related components. Dedicated cell buses have now been replaced by Ethernet-based protocols.
- **Fieldbus (local bus)** This is used to connect process-related components, such as sensors or actuators, and process-related embedded systems with the higher-level process computers.
- **Sensor-actuator bus** This provides a simple and cost-effective way of connecting the components of an embedded system at the process level.

An assignment of the different bus systems to their possible areas of application is shown in Fig. 11.2.

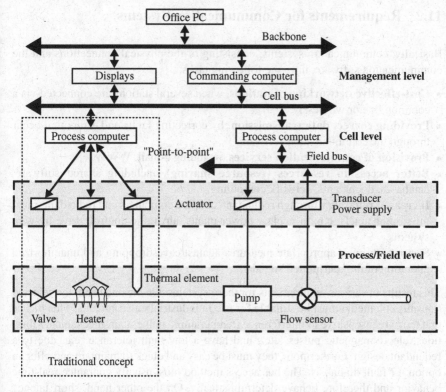

Fig. 11.1 Structure of digital automation systems: usually multiple-layer structure, variations according to the area of application

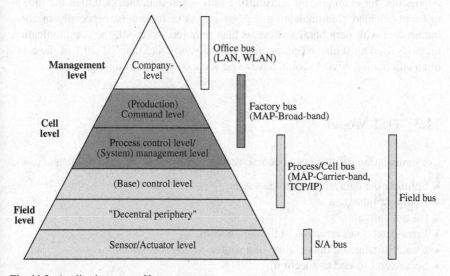

Fig. 11.2 Application areas of bus systems

11.2   Requirements for Communication Systems

Basically, communication systems, consisting of the physical connections and the communication software, have the following tasks:

- **Cost-effective networking of stations**, when several stations are connected via a common line or wirelessly.
- **Providing correct data transmission**, by correcting faults and errors that occur through the communication software.
- **Provision of communication services**, including e-mail, WWW, . . .
- **Better access to resources (resource sharing)**, including shared software, databases, documents, network computing.
- **Increased reliability** through redundant, coupled computers and redundant transmission channels, e.g. in nuclear power plants, air traffic control, drive-by-wire systems.
- **Security** through appropriate measures against eavesdropping and unauthorized alteration of the transmitted signals.

This results in various requirements for these systems. On the one hand, the selected transmission media must be suitable for use in industrial environments. This means that they are insensitive to interference (temperature, ambient air, mechanical vibrations, electromagnetic pulses, etc.) and have a low fault tolerance, e.g. due to a redundant design. Furthermore, they must be easy and quick to maintain and offer an option for fault diagnosis. The bus access method must usually guarantee real-time behavior and therefore behave deterministically. On the other hand, short latency times should occur during communication (more important than throughput), which is possible, for example, by transmitting only small data blocks. Often the option of event-oriented communication, e.g. by means of interrupts, especially in communication with peripheral sensors, is also required. After all, the communication methods used must also be economical. Usually, no more than 10–20% of the costs of an automation device should have to be spent on the bus coupling.

11.3   OSI Model

Communication software can become very complex. This is due to its diverse tasks:

- Splitting the data stream into packages,
- Routing of the data,
- Bit transmission,
- Error detection/correction in the data stream,
- Establishment and dismantling of a source-to-target connection,
- Secure end-to-end connection,
- Encoding/decoding of user data.

Table 11.1 OSI reference model for layers

Layer	Name	Tasks
7	Application layer	Determination of the service of the communication partner for the corresponding applications (e.g. file transmission, email, ...)
6	Presentation layer	Determination of the structure of user data (data format) incl. formatting, encryption, punctuation etc.
5	Session layer	Establishment and dismantling of logical channels on the physical transport system
4	Transport layer	Regulation of the data stream by providing logical channels on the physical transport system
3	Network layer	Determination of routes through the network for a data stream
2	Data link layer	Ensuring of a correct data stream by determination of channel coding, block checking and network access routine
1	Physical layer	Determination of the transmission medium (electrical and mechanical properties)

Table 11.2 Mapping of the layers to the OSI model

Application layer			Manufacturing		Profinet
Presentation Layer	Telnet	FTP	Automation	Profinet	
Session Layer			Protocol (MAP)		
Transport Layer	Transmission Control Protocol (TCP)		User Datagram Protocol (UDP)		
Network Layer	Internet Protocol (IP)				
Data Link Layer	IEEE 802.2: LLC (Logical Link Control)				Profi-
	MAC: Medium Access Control				
	802.3	802.5		802.11	
Physical Layer	Access to the physical medium				bus
	Ethernet	Token ring		WiFi	

Layer-like, hierarchical communication protocols make it possible to master this multitude of tasks. Each layer performs exactly one of the services mentioned. The OSI basic reference model (OSI: Open Systems Interconnection) meets the 7 reference layers defined by ISO as shown in Tables 11.1 and 11.2. Not all of these layers must be explicitly present. For example, often only layers 1, 2 and 7 are found, or several layers are combined within a protocol, as can be seen in Table 11.1.

Only the two lower levels, the physical layer and the link layer, are introduced below, since they form the basis of communication in embedded systems. For further information, see e.g. [1] or [2].

11.4 Physical Layer

The physical layer is used for the transmission of simple bit sequences without any frame. This layer must ensure that logical ones and zeros arrive at the receiver as such, as long as no interference occurs. The treatment of distortions due to interference is only carried out on layer 2. The voltage levels of the logical values and the duration of a transmission step (baud rate), for example, must be specified on the lowest layer. Therefore, the electrical properties of the transmission channel play an important role here. However, mechanical aspects such as the medium and connector located under the physical layer must also be taken into account.

11.4.1 Bandwidth and Maximum Transmission Frequency

Table 11.3 shows the transmission of the character/byte shown in Fig. 11.3 as a binary square wave signal with different transmission rates. The example makes it clear that a minimum bandwidth is required for a transmission channel so that a receiver can detect a transmitted binary signal. Conversely, with a given channel bandwidth, the step size (T_b) of a digital signal must not fall below a limit value. This is the only way to transmit enough harmonics to ensure that the square wave signal is not distorted too much. The waveforms in Table 11.3 show that about ten harmonics are sufficient to recognize the 8 bit character. The step size T_b must then be no less than a quarter of a millisecond (3.33 ms for 8 bit). The values are calculated as follows. If you want to transmit $n = 80$ harmonic oscillations (i.e. a very clean square wave signal) at a transmission frequency of $f_C = 3\,\text{kHz}$, the frequency of the first harmonic oscillation is $f_0 = \frac{f_C}{n} = 37.5\,\text{Hz}$. So you have $T_B = \frac{1}{37.5\,\text{Hz}} = 26.67\,\text{ms}$ of

Table 11.3 Signal distortion caused by a low pass filter with a cut-off frequency of 3 kHz for signals with different speed. The two lowermost signals contain no more information

bps	T_B (ms)	f_0 (Hz) (1. harmonic)	Transmitted harmonics	Respective appearance of the output signal
300	26.67	37.5	80	
2,400	2.33	300	10	
4,800	1.67	600	5	
9,600	0.83	1200	2	
19,200	0.42	2400	1	

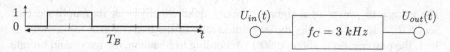

Fig. 11.3 Example of a byte to be transmitted and block diagram of the transmission path

transmission time per byte to be transmitted. This achieves a transmission speed of $\frac{8}{26.67\,\text{ms}}$ bps $= 300$ bps. Or vice versa: If we want to transmit with a speed of 2400 bps, a lower cut-off frequency results as the 1st harmonic to $f_0 = \frac{2400}{8}$ Hz $= 300$ Hz. Thus we can transmit a total of $n = \frac{3\,\text{kHz}}{300\,\text{Hz}} = 10$ harmonic oscillations with a given carrier frequency of $f_C = 3$ kHz, whereby the signal is already clearly beginning to lose its shape, but is still recognizable.

However, this calculation only applies to the transmission of binary square-wave signals. The question arises as to whether there might be another type of coding or transmission of the bit pattern that makes better use of the bandwidth of the channel. If this is the case, the transmitter can encode (modulate) the bit pattern accordingly and the receiver can decode (demodulate) it again.

In general, there is a maximum transmission rate in a band-limited transmission channel that cannot be exceeded even by perfect coding. The existence of such a physical limit for a noise-free channel was published by H. Nyquist in 1924. At that time he had worked out an equation which describes the relationship between the cut-off frequency and the maximum data rate of a channel. In principle, this is the reversal of the sampling theorem (see Sect. 7.3.1).

The sampling theorem states that an analog signal of the bandwidth B can be clearly described and restored by (more than) $2 \cdot B$ samples per second. More samples are unnecessary or redundant. The additional samples correspond to higher frequencies that are not present in the analog signal. If this correlation is reversed, a signal of the bandwidth B can encode a maximum of $2 \cdot B$ independent samples per second. If the sampled values are discrete values with V levels. For the maximum possible data rate MDR, the following applies.

$$MDR = 2B \cdot log_2 V\,[\text{bit/s}] \tag{11.1}$$

The example in Table 11.3 can transmit a maximum of 6.000 bits per second at 3 kHz if the transmission is based on a binary signal ($V = 2$ levels). This is significantly more than is possible with a direct transmission of the binary square wave signal. Shannon later extended this law to include channels with noise. For this purpose, he defined the term *signal-to-noise ratio* (*SNR*) as the ratio of signal power S to noise power N. This ratio is usually described logarithmically ($10 \cdot log_{10} S/N$) with the unit decibel in order to have meaningful, manageable quantities. According to Shannon, the following applies to channels with noise with a bandwidth B.

$$MDR = B \cdot log_2(1 + S/N)\,[\text{bit/s}] \tag{11.2}$$

The energy or power of a signal increases quadratically with its amplitude (see Sect. 5.2.3). If, for example, the amplitude ratio of the useful signal to the noise is 30:1, the power ratio is about 1000:1. According to Shannon, the maximum bit rate depends logarithmically on this power ratio, so the above formula gives a bit rate of about ten times the channel bandwidth. With a bandwidth of about 3 kHz for a telephone line, a theoretical transmission rate of about 30 kBit/s is achieved. Signal coding and modulation have not yet been taken into account, though.

11.4.2   Interference Suppression

Interference on lines can lead to distortion of the signal to be transmitted. Important influencing variables are dispersion (attenuation), reflection, interference with radio transmission and crosstalk (radiation).

Dispersion is the frequency-dependent varied wave propagation of a signal. Since all practical signals cover a whole frequency band, but the individual elementary signals of different frequencies are transmitted at different speeds, the signal is blurred during transmission. The dispersion (and attenuation) of the line on the receiver side turns an ideal square wave signal on the transmitter side into a blurred signal $U_a(t)$ (see Fig. 11.4). This is comparable to the effect of a low-pass filter.

Depending on the structure of a cable, it is more or less dispersive. High-frequency cables used in practice are dispersion-free, while the double-wire conductors and coaxial cables used for data transmission are dispersive. To prevent the dispersion from becoming too large, amplifiers must be installed at regular intervals on longer lines to reprocess the blurred signal.

A *reflection* is the disturbance of the signal due to mismatching of the characteristic impedance of the line. If a signal wave is transmitted, the signal is reflected at the end of the line with the reflection factor r.

$$r = \frac{R_T - Z_W}{R_T + Z_W} \tag{11.3}$$

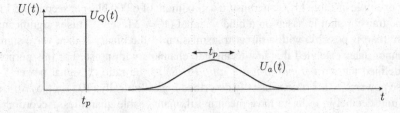

Fig. 11.4 Example for dispersion

Fig. 11.5 Symmetrical/differential transmission

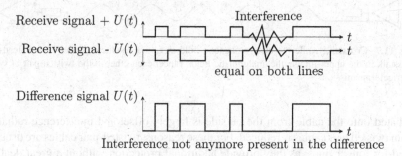

Fig. 11.6 Influence of interference pulses on a differential transmission line

Here, R_T designates the terminating resistance and Z_W the characteristic impedance of the cable. If $r = 0$, the wave characteristic is correctly adjusted, otherwise the wave is reflected with the result that incoming and reflected wave overlap. This fault can therefore be avoided if $R_T = Z_W$, which can be achieved with an appropriate terminating resistor.

In order to avoid errors due to interference pulses, instead of measuring the voltage of one conductor against a common ground potential for each signal (asymmetrical transmission), each signal can be transmitted via two lines with opposite voltages (against ground). This is referred to as symmetrical or differential transmission (Fig. 11.5).

The differential transmission increases the signal-to-noise ratio and thus allows longer transmission distances. In Fig. 11.6, the opposite voltages of the lines "+" and "−" to the ground and the differential voltage encoding the signal are shown. An external interference with the signals is shown in the middle of the figure. The interference, whether capacitive or inductive, affects both conductors in the same way. If the signal were merely described—as in asymmetrical transmission—by the voltage on a line to ground, the information on the receiving side would be incorrect due to the interference. Since, however, in the symmetrical case, the signal is expressed by the voltage difference and the interference is the same on both lines, there is no interference in the signal on the receiving side (even if the interference on the individual lines can be considerable). Clearly, therefore, the symmetrical design can considerably increase the signal-to-noise ratio.

The twisting of pairs of conductors with symmetrical transmission (Fig. 11.5) causes the electric fields to be largely extinguished (Fig. 11.7) because the conductors are intertwined with one another. This technique is called a *twisted pair*. Interference

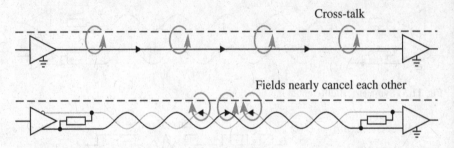

Fig. 11.7 Crosstalk: while the electromagnetic field around a conductor under current leads to crosstalk in the upper image, the antagonistic fields cancel each other in the twisted pair at symmetrical transmission

radiated onto the cable from the outside is largely offset and interference radiated from twisted pair cables is reduced. For these reasons, twisted-pair cables are usually used for longer lines, as they provide additional protection without a great deal of extra work. The higher the transmission rate, the tighter the twisting of the cable pairs.

11.4.3 Clock Recovery

When transmitting data, it is important to synchronize the clock of the transmitter and the receiver. This is the only way for the receiver to know when to expect a new bit. Otherwise, for example, three consecutive 'one's could only be recognized as one or 5 bits. Many bus standards therefore provide for an external clock line. This procedure is very accurate and relatively interference-free. However, the additional cable can be a problem, especially in systems with limited space or cable harnesses that are already overflowing. For serial data transmission without a clock line, the clock must be obtained from the data signal. The signal edges, i.e. the change of the signal level (1-0, 0-1), are used for this purpose, since these take place almost synchronously with the corresponding signal edges of the clock signal. Thus the receiver can synchronize its internal clock with the clock of the transmitter at these points. If there are no level changes for a long time, the clock of the receiver may deviate increasingly from that of the transmitter. Frequent flank changes are therefore required for clean clock recovery. Possible ways to force this are the use of a scrambler, which converts the data stream, and line codes. Two ways of setting up a scrambler are *bit stuffing* and creating a *pseudo-random sequence*.

With bit stuffing, the transmitter always inserts a zero/one into the data stream after a sequence of fixed lengths n of ones and zeros. The receiver then interprets each $111\ldots110$ sequence as $111\ldots111$ (and vice versa).

The following example illustrates the procedure. After $n = 5$ ones, a zero is inserted into the bit sequence (see Ethernet).

Example: n = 5: 0011111̱1100011001111100 (Original signal)

00111110̱110001100111110̱00 (Signal with bit stuffing)

The alternative approach to bit stuffing is to generate a pseudo-random sequence from which the original data stream can be uniquely recovered. These random sequences guarantee a frequent edge change without increasing the signal length as with bit stuffing. The pseudo-random sequences are generated by scramblers based on feedback shift registers. The descramblers are constructed symmetrically to it.

If these line codes are suitable for recovering the clock from the signal, they are *self-clocking line codes*. In addition to clock recovery, these codes can also meet other requirements. *Line codes without a DC component in the signal*, for example, allow data to be transmitted via a power supply line if a DC voltage is used as the power supply. In this case, the alternating signal of the message is applied to the DC voltage. On the receiving side, the communication partner can use a suitable filter to separate the data signal and the direct component from the power supply. Sometimes DC-free codes are also used where the specification prescribes an AC voltage for physical or other reasons. For example, to ensure compatibility with old analog devices, Telekom also prohibits the use of codes with DC components in digital telecommunications (ISDN, DSL).

Table 11.4 shows a sample of frequently used line codes.

Table 11.4 Examples of line codes

Clock Data	1 0 0 1 1 0 0 0 0	Comment
NRZ ("Non-Return-to-Zero")		Single bandwidth, not free of DC, no clock
RZ ("Return-to-Zero")		Double bandwidth, long sequences of zeroes, no clock
Bipolar. code		Long sequences of zeroes, no clock, free of DC
Manchester (e.g. Ethernet)		Clock recovery, double bandwidth
4B/5B-Code 4-bit block → 5-bit codeword (e.g. FDDI)	0000 → 11110 0001 → 01001 0010 → 10100 0011 → 10101 0100 → 01010 0101 → 01011 0110 → 01110 0111 → 01111 1000 → 10010 1001 → 10011 1010 → 10110 1011 → 10111 1100 → 11010 1101 → 11011 1110 → 11100 1111 → 11101	Clock recovery, 25% increased bandwidth

11.5   Link Layer

The roles of the protocols on the link layer are the transmission of *frames* (several 100 bytes) between two neighboring stations, which are passed on to the physical layer, including handling of transmission errors. For this purpose, the frames formed by the link layer are framed by special start and end sequences (*header* and *trailer*). As a rule, the receiver is responsible for error detection and acknowledges each received frame accordingly.

For buses with several connected stations, the receiving station must also be addressed. If several potential transmitters are connected to a common bus, signal collisions occur on the line if two or more stations transmit simultaneously. To prevent this, the stations receive an arbitration component; in the network area this is referred to as *medium access control* (*MAC*).

A protocol on the link layer takes account of the following aspects:

- Definition of header and trailer (e.g. receiver address, CRC check polynomial, . . .),
- Adaptation of different read/write speeds of the communication participants to prevent buffer overflows (flow control),
- Error detection with definition of error-detecting codes, receipts to avoid falsification and loss, timers to avoid frame, receipt or token loss and sequence numbers to avoid duplicates,
- Error handling with definition of error correcting codes, frame repetitions, sorting of duplicated and falsified frames.

11.5.1   Error Protection

No matter how hard you try to reduce interference with countermeasures at the analog level, you will never achieve 100% secure signal transmission. Although technical measures such as shielding, potential separation, etc. help to reduce the probability of bit errors, they cannot guarantee correct transmission. There will always be signal distortions which lead to errors in the transmission of digital signals due to changes in the bit pattern (bit errors). Such errors often occur more frequently during one period of time (error bursts). The error rate generally increases with the transmission speed. Nowadays the residual error probability can be reduced to such an extent, however, that transmission can practically be regarded as secure. This is achieved by additional software measures, whereby a received message is checked for errors and in the event of an error, appropriate action is taken. Securing the data in this way always consists of the two steps of *error detection* and *error correction*.

In practice, error detection plays a more important role than automatic error correction using error-correcting codes. If the bit error probability is low, it is much more efficient to repeat a message recognized as faulty (ARQ, automatic repeat request) than to use complex error-correcting codes. In the case of highly disturbed channels such as radio transmission, efficient error-correcting codes (e.g. Hamming code, convolutional codes) are used, which generally do not guarantee completely correct transmission, but reduce the residual errors to such an extent that the ARQ procedure mentioned above can be used without blocking the medium with continuous message repetitions.

Since the receiver usually does not know in advance which data it will receive next, the prerequisite for error detection is redundant data. However, since it would be time-consuming to transfer all data several times, a trick is used. To do this, the sender divides a data stream into a sequence of data blocks and calculates redundant check information of length r bit for each data/payload block of length k bit. Payload and test information are then sent to the recipient, who calculates the test information again after receiving the data. If the recalculated test information does not match the received test information, it is assumed that an error occurred during transmission.

The total length of the transferred data block is $n = k + r$ bit. The redundancy introduced with the test information is r/k. In principle, the probability of error detection increases with the number of check bits. On the other hand, it should not be forgotten that the transmitted test data itself can also be falsified, i.e. the probability of error increases with the length of the test data. In practice, a compromise in the length of the test data is therefore necessary which must also be cost-effective.

Today, there are mainly two approaches to error protection:

- parity check and
- cyclic redundancy check (CRC).

During parity checking, a check bit is added to each data block of length k bit. This check bit is assigned in such a way that the number of all ones in the extended data block is even or odd. The redundancy for a parity bit is $1/k$. With such a 1-bit parity, any odd number of falsified bits, especially all 1-bit errors, can be detected in the block. 2-bit errors and other error bundles, for example, are not recognized.

However, parity checking is not used for serial data transmission over networks—be it fieldbuses, LANs or wireless networks—because error bundles are the rule rather than the exception here. The problem cannot be solved in principle by increasing the number of parity bits. Computer networks therefore use a method that is mathematically more complex but easy to implement, the CRC test. The CRC procedure adequately secures data blocks from a few bytes up to several kilobytes in length with 8 to 32 check bits.

Another approach is echo control. The receiver sends the data block back to the sender, who checks this "echo". This procedure is useful for ring structures.

The CRC process is based on polynomial division. The approach is to interpret the message to be transmitted as a binary polynomial and to divide it in the transmitter by a fixed test polynomial (*generator polynomial*). The resulting remainder is then attached to the message as a trailer. If the receiver comes to the same remainder with the same arithmetic operation, the transfer was successful.

From a mathematical point of view, the bits of a bit sequence $b_{n-1}b_{n-2} \ldots b_1 b_0$ serve as coefficients of a polynomial $P(x)$ of the degree $n - 1$.

$$P(x) = b_{n-1}x^{n-1} + b_{n-2}x^{n-2} + \cdots + b_1 x + b_0$$

For example, a polynomial division looks like this

$$(x^3 + x + 1)/(x + 1) = x^2 - x + 2 \text{ Remainder } -1$$
$$\underline{-(x^3 + x^2)} \quad \{= x^2(x + 1)\}$$
$$-x^2 + x + 1$$
$$\underline{-((-x^2) - x)} \quad \{= (-x)(x + 1)\}$$
$$2x + 1$$
$$\underline{-(2x + 2)} \quad \{= 2(x + 1)\}$$
$$-1 \quad \{\text{Remainder}\}$$

The remainder is of interest for the CRC procedure. Now the data transfer is considered using the CRC procedure. If

- **N(x)** is the message polynomial of degree $(k - 1)$,
- **G(x)** is the generator polynomial of degree r (same on transmitter and receiver side),
- **D(x)** is the remainder of division $N(x) \cdot x^r / G(x) \rightarrow$ degree $< r$ (CRC field, typically 16/32 bits),
- **S(x)** $= N(x) \cdot x^r + D(x)$ is the transmitted code polynomial of degree $n = k - 1 + r$,
- **E(x)** is the error polynomial due to transmission error (degree $\leq n$) and
- **R(x)** $= S(x) + E(x)$: is the received code polynomial.

The following applies to the transmission.

$$S(x) = N(x) \cdot x^r + D(x)$$
$$R(x) = N(x) \cdot x^r + D(x) + E(x)$$

Payload data and the remainder $D(x)$ of the division $N(x) \cdot x^r / G(x)$ are sent. The following calculation is carried out on the receiving side.

$$\frac{R(x)}{G(x)} = \frac{N(x) \cdot x^r + D(x)(+E(x))}{G(x)}$$

A correct transmission took place if $R(x)$ is an integer multiple of $G(x)$, i.e. the remainder of the division $R(x)/G(x)$ is zero.

With a 16-bit test polynomial used in practice, all odd bit errors and 2-bit errors for blocks up to 4095 bytes and all error bursts up to 16 bits in length and 99.997% of all longer error bursts are detected. In addition, the CRC calculation can be implemented easily in the hardware using an extended shift register. An example of the CRC calculation can be seen in Fig. 11.8.

The message 110101 is to be transmitted here. The generator polynomial is 101. The resulting remainder is 11. Note that the binary polynomial division is applied here. This corresponds to an XOR operation of the individual bits, which in practice

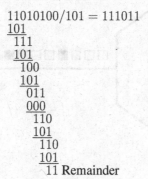

$$11010100/101 = 111011$$

$$\underline{101}$$
$$111$$
$$\underline{101}$$
$$100$$
$$\underline{101}$$
$$011$$
$$\underline{000}$$
$$110$$
$$\underline{101}$$
$$110$$
$$\underline{101}$$
$$11 \text{ Remainder}$$

Fig. 11.8 Example for the calculation of the CRC polynomial

increases the efficiency of the algorithm compared to a "real" division. Thus the sent message results in 11010111.

11.5.2 Media Access Control

If several participants want to use the same medium, signal collisions can easily occur (as shown in Fig. 11.9). *Medium access control* (*MAC*) is therefore required to determine who can use the medium and when. With serial transmission media, this access control should be as deterministic as possible in order to enable fair or priority-oriented distribution.

One way to control access is *fixed assignment types* such as frequency division multiplexing and time division multiplexing. In frequency division multiplexing, fixed carrier frequencies or channels are assigned (see radio, TV), in time division multiplexing, the subdivision into fixed time windows takes place as shown schematically in Fig. 11.10 using the example of TDMA (time division multiple access).

Characteristic of frequency division multiplexing and time division multiplexing are the real-time guarantee; they are static concepts, therefore the number of transmitters and data volume must be known. Fixed assignment is useful for regular data transfer, but is not recommended due to the poor load with fluctuating data volume.

Fig. 11.9 Initial problem of media access control

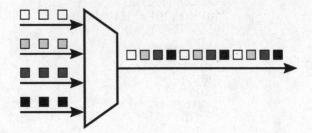

Fig. 11.10 TDMA

An alternative is offered by external control lines which, independently of the bus, are only used to control or negotiate media access. This can be used to implement various procedures. One example is the central procurement procedure using dedicated request and grant lines (see Fig. 11.11). Here, all stations willing to transmit send a request to a central arbiter, which then controls media access on the basis of an allocation procedure implemented there.

A decentralized procedure is polling via control lines, in which the access mechanism is implemented redundantly in all stations (see Fig. 11.12).

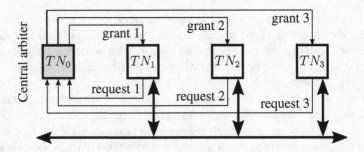

Fig. 11.11 Bus arbitration from a central arbiter (special control lines for parallel busses)

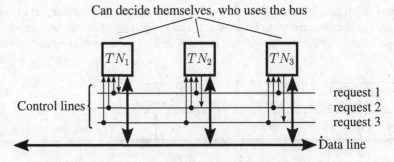

Fig. 11.12 Distributed bus arbitration (additional control lines for parallel busses)

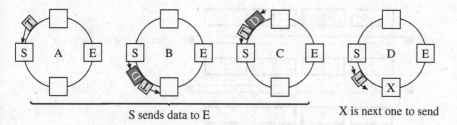

S sends data to E   X is next one to send

Fig. 11.13 Example of a transmission in a token ring

The major disadvantage of this method is that control lines must be laid in addition to the actual bus. This can be particularly problematic in large networks or in systems with strict weight requirements. These methods can also only be extended to a limited extent, since the physical possibility of connecting further lines is often limited, e.g. by the number of pins or the installation space. In these cases, procedures that do not require a change in bus topology are advantageous.

The *master/slave method* is characterized by central bus assignment. Only the master has the right to initiate a communication cycle, other devices (slaves) send only at the request of the master. This ensures that never more than one device transmits at any one time. Most frequently used is the polling procedure, in which the master operates all slaves one after the other. Direct communication between slaves is not possible. Slaves require little effort, since complete implementation of the protocol is not necessary.

With the procedure involving a *token ring/token bus*, a token represents the sending authorization. The participant with a token passes it on to the next participant as soon as there is no longer a need for communication or after a period of time has elapsed. Figure 11.13 shows the sending process.

This is generally a fair access procedure, priority allocation is possible by adding priority information to the token. With maximum data block length, a real-time guarantee is possible. A special case (e.g. PROFIBUS) has active (master) and passive participants (slaves), whereby the slaves send only after direct request by a master (communication via master/slave protocol) and masters communicate with each other via token passing.

Since there is a maximum token circulation time, real-time capability is guaranteed. Due to the maximum token circulation time, the loss of the token can also be detected and defective participants can be ignored during token assignment (logical ring). The cycle time is reduced by the master/slave mechanism. However, the token rotation time increases with the number of (active) participants, and there is also a great deal of reconfiguration effort if the number of participants changes. Another disadvantage is that passive stations cannot transmit acyclically (e.g. alarm message).

CSMA (carrier sense multiple access) is MAC with random access. There is no exchange of control information, so collision of messages is possible—the collision can falsify sent data. These can no longer be reconstructed. To avoid collisions, the medium is monitored before transmission and transmission takes place only if it is

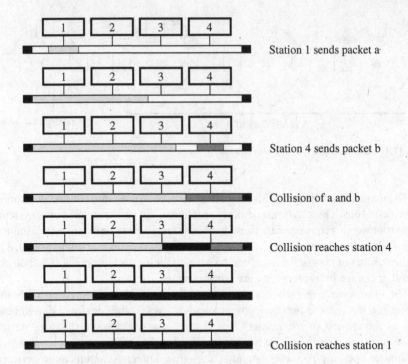

Fig. 11.14 Time shift during a transmission with CSMA

free. Nevertheless, collisions are still possible when a channel becomes free, e.g. due to the transit time of signals in the line (see Fig. 11.14). *CSMA/CD* (collision detection: termination on collision detection, e.g. Ethernet) and *CSMA/CA* (collision avoidance: direct detection of collisions, often in the fieldbus area, e.g. CAN) are examples for this.

With CSMA/CD, the transmitter reads on the channel and checks that the transmitted signal matches the read signal. If a collision is detected, an interference signal (jam) is transmitted and the transmission is stopped. A collision is detected at the latest after twice the maximum signal propagation time t_S (lapidary: if the signal has passed through the bus once and returned again, see Fig. 11.15), therefore the minimum packet length must be $2 \cdot t_S$ and a maximum cable length must be fixed. Collision detection cannot guarantee real-time capability, because arbitrarily long delays can occur, e.g. if a large number of stations want to transmit and therefore many consecutive collisions occur. Efficiency therefore drops sharply as the load increases.

However, CSMA/CD allows a high to very high number of subscribers at a low transmission rate, and these can also be connected or removed without a great deal of configuration effort. Since transmissions only take place when required, there is no (temporal) overhead.

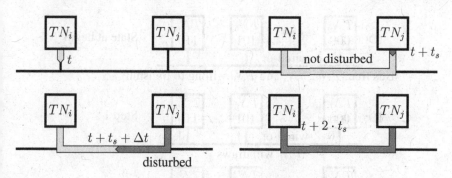

Fig. 11.15 Determining the maximum time until a collision is detected

CSMA/CA aims to avoid collisions. A station willing to transmit reads the channel. If it is blocked by another communication, the station waits for a random period of time, then checks again. If the channel is free, it transmits it's data. The period of time is often limited to certain minimum and maximum waiting times and can also be altered under heavy communication load on the channel.

This approach is usually extended by additional measures. Such a measure is CSMA/CR (collision resolution). It additionally resolves collisions by priority-controlled bus allocation. Each participant has an ID which corresponds to the priority. At the end of a current bus transmission, all subscribers willing to transmit begin to transmit their ID, whereby the subscribers are wired-OR-linked ("1" dominant, "0" recessive) or wired-AND-linked ("0" dominant, "1" recessive). The transmission begins with the highest value bit. If the current identifier bit on the bus is not equal to the one sent by the user, the recessive participant withdraws, as shown in Fig. 11.16. The prerequisite for this is that the signal propagation time t_s is negligibly small compared to the step size (bit time) t_B. $[t_s = \frac{l}{v}] << [t_B = \frac{1}{R}]$, with l = Line length, v = Propagation speed, R = Transmission rate; e.g. $v = 0.66 \cdot c$, $R = 10$ MBd $\Rightarrow l << \frac{v}{R} \Rightarrow l << 20\,m$

If the packet length is limited, the participant with the highest priority has real-time behavior. The bus can be blocked if the subscriber with the highest priority always sends, therefore it is generally necessary to wait a certain time before the next transmission request after a data transmission (requirement for application layer). If the waiting time is long enough, further participants can also become real-time capable.

There is no overhead here either, as transmissions only take place when required. Compared to collision detection, the bus can be used optimally because a message always gets through. In addition, high-priority messages are transmitted without delay. Since the signal propagation time is small compared to the transmission time per bit, possible collisions can only be detected during quasi-simultaneous transmission. The signal propagation time limits the product of line length and transmission rate, so the transmission rate breaks down at long line lengths.

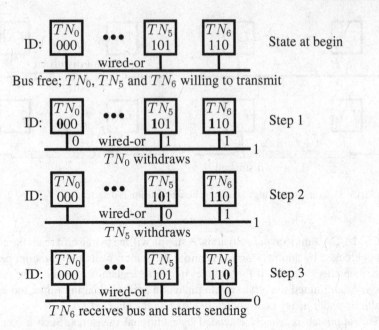

Fig. 11.16 Example for CSMA/CR, using wired-OR

11.6 Examples of Commercial Bus Systems

Due to the widely differing requirements of the individual areas of application, it is difficult to meet the demand for a universally applicable communication system. Since the universal fieldbus is not feasible, manufacturers are trying to establish their own systems on the market, leading to fragmentation and system diversity.

The international fieldbus standard, which was introduced in 1984, has practically failed as a result. As an accepted solution, de facto standards (e.g. Profibus, CAN) were combined in a new standard EN-50170.

11.6.1 Sensor-Actuator Buses

Sensor-actuator buses are closest to the process level. Here the communication systems connect robots, numerical controls (CNC), programmable logic controllers (PLC) etc. in an industrial environment with each other and with their sensors and actuators. Important aspects of the communication media in this area are the real-time behavior mentioned above (fixed system reaction times in the millisecond range are required), use under field conditions and also cost-effectiveness.

Most sensor-actuator solutions are manufacturer-specific. Industrial examples are the actuator-sensor interface ASi, the VariNet-2 and the Interbus-S. They all have in

common the fact that the bus access—in contrast to fieldbuses—takes place using the
very efficient master/slave method. The central controller or the higher-level fieldbus
is the bus master that interrogates the sensors and actuators (slaves).

ASi is a two-wire bus system for direct coupling of binary sensors and actuators
via a bus with a higher-level controller. The higher-level unit can be a PLC, a CNC,
a microprocessor or PC, or a gateway to a higher-level fieldbus. The aim of ASi is
to replace the wiring harnesses at the lowest level which still predominate.

ASi was originally developed by 11 manufacturers of sensors and actuators. In
addition to the general requirements for sensor-actuator buses, the specifications for
the development of ASi were also given: two-wire cable, transmission of data and
power for all sensors and most actuators via the bus, straightforward and robust
transmission method without restrictions regarding network topology, master-slave
concept with one master and a small, compact and cheap bus connection.

The master-slave procedure of an ASi system corresponds to a token ring with
only one active participant. The ASi slave chip can either be installed in a separate bus
coupling module to which the sensors and actuators are connected in the conventional
way, or it can be installed directly in the sensor/actuator. The latter is ASi's long-
term approach, while the first variant is important in order to continue using today's
sensors and actuators. The sensors and some of the actuators are powered via the bus.
Actuators with higher power consumption (more than 100 mA at a standard supply
voltage of 24 V) require their own power supply.

As with normal electrical installations, the topology of the network can be arbitrary: linear, with stub lines or branched like a tree. The bus ends do not need to be
terminated, but the maximum cable length is limited to 100 m. Larger distances must
be bridged by repeaters. Branches in the network are made with coupling modules
that passively connect two lines to each other. Two alternative, unshielded two-wire
lines are specified as the transmission medium. The power supply unit and the cables
are designed for a maximum of 2 A at 24 V in both cases. The two cable variants are

- 1.5 mm^2 ribbon cable/flat-band power cable (inexpensive),
- 1.5 mm^2 ASi-specific flat ribbon cable (has advantages during installation—cannot
  be connected incorrectly due to special cable geometry).

With ASi, only one master and 31 slaves with a total of 124 sensors/actuators are
permitted per bus or strand. However, several ASi strands can be connected in parallel.

In the ASi flat cable coupling module, contact is made by a penetration technique.
The installation is carried out simply by clipping the ASi cable into the coupling
module (without cutting and stripping). Each coupling module can accommodate
two cables and connect them electrically, thus creating the branches. The user module
is located in the cover of the coupling module. User modules are available in a wide
variety of forms, but they can be divided into two classes:

- **active user modules** containing the electronics of the actual slave interface—up to
  four conventional sensors/actuators can be connected to these units,
- **passive user modules** without their own electronics—these are only used for
  further branching of the ASi cable.

The components of the slave are integrated in an IC. No processor or software is required to connect the sensors/actuators to the ASi bus via these slave connections. All frame processing is done in the IC.

When defining the modulation method, care was taken to ensure that the transmission signal was DC-free and narrow-band. DC-free operation is necessary so that the data signal and the power supply can be superimposed. Narrow banding is required, as the attenuation of the cable increases rapidly with frequency. In addition to these two main requirements, care should be taken to ensure that the signal is easy to generate.

In view of these requirements, the ASi consortium decided to use alternating pulse modulation. The raw data are first Manchester-coded. This results in a phase change each time the transmit signal changes. A transmission current is then generated from the Manchester code. This transmission current induces a signal voltage level, which can be greater than the supply voltage of the transmitter, via an inductance present only once in the system. When the transmission current rises, there is a negative voltage in the cable and when the current drops, there is a positive voltage. On the ASi lines, bit times of 6 μs, i.e. a transmission rate of 167 kbit/s, can be realized by means of alternating pulse modulation.

On the protocol level ASi uses a simple bus access procedure via cyclical polling by the master. Interrupting is not intended. Under certain conditions, this procedure can be used in real time. The slaves are addressed cyclically by the master always in the same order. The master sends a frame with the address of a slave (14 bit at 6 μs), to which the slave must respond within a specified time (7 bit at 6 μs). The time between the master call and the response by the slave is called the master pause. It is generally three, but a maximum of 10 bit times. After that, the master assumes that there is no answer and sends the next request. The slave pause, i.e. the time between the slave response and the following master call, is a bit time of 6 μs. With only 5 bit, the frames contain very short information fields to keep the messages and the bus assignment by individual sensors/actuators short. If all fields of a cycle are added up ($14 + 3 + 7 + 1$ bits), the result is 150 μs per cycle or 5 ms total cycle time for 31 slaves and 20 ms for the maximum configuration level of 124 sensors/actuators. These times are sufficient for PLCs. The real-time capability for a given application depends on the number of sensors/actuators.

The security of data transmission is of interest for ASi. The check of the received data is carried out according to different criteria than in the bus systems considered so far, as the frame of the checksum overhead would be too large because of the brevity.

With ASi, the signal curve is tested on the physical layer. For this purpose, the received signal is sampled sixteen times during a bit time and the following set of rules is continuously evaluated by the slave module in real time:

- with start/stop bits the first pulse must be negative, the last pulse positive,
- successive pulses must have different polarity,
- only one pulse may be missing between two pulses in one frame,
- no pulse in pauses,

- even parity.

This rule set in itself results in a high degree of security. Studies have shown that all single and double pulse errors and 99.9999% of all triple and quadruple pulse errors are detected. In fact, the parity check is only effective from triple pulse errors. Theoretical estimates show that statistically speaking, at a bit error rate of 100 errors/s a faulty frame is missed only every 10 years.

Faulty frames are repeated with ASi. Due to their brevity, however, they hardly increase the overall cycle time at all.

11.6.2 Fieldbuses

Fieldbuses are located above the sensor/actuator level, although the two levels cannot always be precisely delimited. In contrast to sensor-actuator buses, fieldbuses are multimaster systems. The resulting bus access protocol (arbitration) makes the bus system more flexible, but also more complex and inefficient.

In addition to the general requirements for buses in automation technology mentioned above, the specifications for fieldbuses also include the following:

- the dimensions are between a few meters and a few kilometers,
- the bus system should be flexible enough for additional bus subscribers to be added without problems,
- hard time requirements with guaranteed maximum reaction time of the system, real-time capability and reaction times in the millisecond to second range (depending on the application).

Two areas of the group of fieldbuses will be discussed with their prevailing bus standards. These are process automation (PROFIBUS) and automotive technology (CAN). A good overview can be found in [2] and [1].

11.6.2.1 Profibus (PROcess FIeld BUS)

The *Profibus* was developed by 14 manufacturers and 5 scientific institutes and is an international fieldbus standard (IEC 61158/IEC 61784). There are three variants of it: Profibus-FMS, Profibus-DP and -PA.

The basic variant *Profibus-FMS* is located on the higher system bus levels due to its low speed. *Profibus-DP* expands the Profibus-FMS into the sensor/actuator system range. Complex sensors and actuators should be able to use the Profibus. *Profibus-PA* is a further extension for the field of process automation that requires intrinsically safe data transmission.

The Profibus is a token bus, more precisely, a multi-master bus with token-passing access procedures. The physical network structure is a line topology (bus) with short

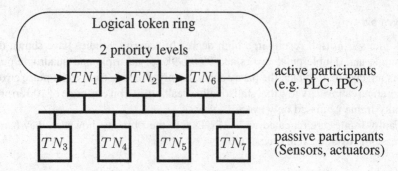

Fig. 11.17 Topology of Profibus

stubs (Fig. 11.17). Profibus-FMS and Profibus-DP can use the same medium at the same time.

The transmission medium for the Profibus is either fiber optic or shielded twisted pair; in the latter case the bus segments must be passively terminated. RS-485 was defined as the bus interface. The line signal results from the NRZ (no-return-to-zero) coding.

The maximum cable length is a few hundred meters, but depends heavily on the transmission speed. The bus can be extended by a maximum of three bidirectional repeaters between two stations. The upper limit of the bus participants is 32 for time-critical applications and 125 stations for applications that are not time critical.

Master and slave stations can be connected to the Profibus. Bus access is a hybrid process. All active participants (master stations, e.g. PLC) participate in bus arbitration according to the token passing procedure. After receiving the token, the current bus master can communicate with any participants, including passive slaves (e.g. sensors/actuators). The master station calls to its partner stations for communication (master-slave communication or polling).

The bus master can carry out or initiate a data exchange once or several times during a cycle, whereby the total communication time depends on various temporal boundary conditions. For this purpose, the target token circulation time specified as a parameter is compared with the measured, actual token circulation time. The bus occupancy time depends on the remaining time reserves. However, each master may send at least one high-priority message. Other "normal" messages, on the other hand, are only permitted if the token target circulation time has not yet been exceeded. By passing on the token at the latest after the hold time derived from the token circulation time, accurately predictable real-time behavior is possible.

Using various mechanisms, the Profibus FMS can detect failed stations and remove them from the token passing ring. If a failed station becomes fully functional again, it is also automatically reinstalled. Faulty frames are rejected by the recipients and must be repeated.

The OSI layers 3 to 7 are not fully developed in Profibus. The application layer sits directly on the link layer on which the frame formats are defined. The application layer is divided into two sublayers:

- **LLI (lower layer interface)** an interface between the application layer and the link layer that monitors the data transfer,
- **FMS (fieldbus message specification)** an interface with the user that provides a variety of services.

The time that elapses before the master receives the information of a slave increases with the number of slaves in the system. The reaction time of the system becomes worse the more slaves are connected. The target token circulation time represents the worst case for a circulation. If one or more masters fail, the system optimizes itself by means of special messages. The system always knows which masters are still active.

The overhead of the Profibus can be considerable. Each character already has a 3-bit overhead (start, parity, stop). In addition, the frames have varying numbers of control characters. In extreme cases, up to 90% control characters are required.

11.6.2.2 CAN (Controller Area Network)

CAN was developed in 1981 by Bosch and Intel with the aim of networking complex controllers and ECUs. CAN was used internationally, mainly in the motor vehicle sector to replace increasingly complex wiring harnesses (up to 2 km, 100 kg), but also in the household appliance sector (Bosch), in textile machinery, in medical equipment and some other applications. Since the bus can also be used in principle as a sensor-actuator bus in compliance with real-time requirements, more and more fields of application such as building automation have recently opened up. An advantage of CAN lies in the low-cost bus coupling components resulting from the large quantities.

CAN is a multi-master bus with serial transmission. The wiring is completed using bus topology. The CAN transmission medium is a shielded twisted pair cable. The differential voltage levels are intended to help prevent interference from electromagnetic radiation. In the event of a fault, communication is also possible via a single-wire line (and common ground). Special switching devices and fault measures then switch over to asymmetrical transmission.

CAN allows a quite high data transmission rate of 10 kbit/s up to 1 Mbit/s with bus lengths of 40 m up to 1 km. In practice, effective bit rates of 500 kbit/s are achieved. In the bus specification, particular emphasis was also placed on transmission reliability and data consistency, since the bus is exposed to severe interference in the automotive sector, for example. By means of various measures, including a 15 bit CRC field, a residual error probability of 10^{-13} is achieved.

Short messages in blocks of max. 8 bytes are transmitted via the CAN bus to enable low latency and short reaction times. A distinction is made between high-priority and normal messages. With 40 m bus length and a transmission rate of 1 Mbit/s the maximum reaction time for high-priority messages is 134 μs. However, it has to be taken into account, that many high-priority messages can block the bus access for messages with lower priority. CSMA/CR is used as the bus access method.

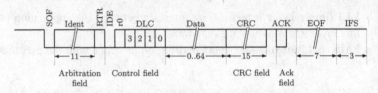

(a) Format of the data frame (CAN 2.0A).

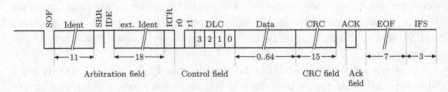

(b) Format of the data frame (CAN 2.0B).

Fig. 11.18 Data frames for different CAN formats

At the beginning of a transmission cycle during the arbitration phase, the station ready to transmit with the highest priority is determined. This is the station with the smallest (source) address. Bit-based, priority-controlled arbitration distinguishes between dominant ("0") and recessive ("1") voltage levels. The prerequisite for this CSMA/CR procedure is of course that all stations start arbitration at the same time and have the same clock speed.

In contrast to most bus protocols, CAN is message- or object-oriented and not station-oriented. This means that the frames do not contain any addresses. Every message is a broadcast message. A station must itself determine from a frame whether the current message, or more precisely the message type, is intended for it. For this purpose, an identifier field (*ID/Ident* or *arbitration field*) is available in the data frame, the value of which generally describes the sender. For the identifier field, 11 Bit are provided in the frame, so that a maximum of 2032 different message objects can be distinguished by the identifiers (16 identifiers are reserved). The typical CAN data frame format is shown in Fig. 11.18a, b. CAN has a total of four different frame formats: data frame, remote frame, error frame and overload frame.

The standard frame consists of:

- **SOF** Start Of Frame
- **EOF** End Of Frame
- **IFS** Inter Frame Space
- **RTR** Remote Transmission Request or **SRR** Substitute Remote Request (request for a message): 0 = Data Frame, 1 = Remote Frame
- **IDE** Identifier Extension bit (CAN2.0A: "0", CAN 2.0B: "1")
- **Control field** r0, r1 reserved for Extended CAN, 4 bits data length (DLC: Data Length Control)
- **ACK** Acknowledge, changed by receiving station.

While data is generally transferred in the data frame, a remote frame can also be created by setting the *RTR* or *SRR* bit, which requests certain data. These are in turn described by the identifier field. For this purpose, the data field must correspond to the size of the expected response data. If a bus station detects an error in the network, it can send an error frame to alert the others. This consists of 6 *ERROR FLAG* bits followed by 8 recessive bits (ones). A distinction is made between the active error frame, where the ERROR FLAGS are dominant bits, and the passive error frame, where they are recessive. Finally, an overload frame can be used to create a forced pause between data frames and remote frames, e.g. when a participant's input buffer is full. It corresponds to an active error frame, but can only occur at certain points of the communication, making it distinguishable from it.

References

1. Barry JR, Lee EA, Messerschmitt DG (2012) Digital communication. Springer Science & Business Media, New York
2. Wolf M (2017) Computers as components: principles of embedded computing system design, 4th edn. Elsevier Morgan Kaufmann, Amsterdam

Part IV
Modeling and Real-Time

Chapter 12
Software Modeling and Analysis

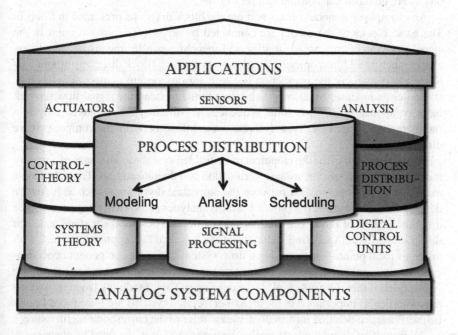

This chapter introduces various methods for modeling embedded systems. In addition to various diagram types for UML, SDL is also considered. This provides tools for graphical modeling that enable interdisciplinary communication and a clear structuring of the development of embedded systems. Finally, Petri nets are introduced which are a frequently used tool for analysis of concurrent processes, as frequently occur in embedded systems.

© The Author(s), under exclusive license to Springer Nature Switzerland AG 2021   319
K. Berns et al., *Technical Foundations of Embedded Systems*, Lecture Notes in Electrical
Engineering 732, https://doi.org/10.1007/978-3-030-65157-2_12

12.1   Abstraction of Technical Systems

Embedded systems are becoming increasingly important in today's world as a result of increasing digitization. Increasingly complex tasks are being taken over by them. At the same time, the complexity of the embedded systems themselves is also increasing. In addition to simple systems consisting of little more than one 8-bit microcontroller, there are also complex control networks with hundreds of often different processing units, integrated smart sensors and corresponding actuators. Another factor is the increased work involved in software development, since a distributed, heterogeneous and often highly parallel system has to be used. In order to master this complexity, various methods for abstract modeling and analysis have been developed. The basic idea is to find a uniform, mostly graphical representational form for the heterogeneous systems which depicts a specific aspect, such as the structure, the signal flows and/or the functional sequences. For this purpose, predefined basic blocks are connected according to a set of rules.

An example of a model of this sort are the block diagrams presented in Chap. 6. The basic blocks of the model are connected by arrows. The arrows point in the signal flow direction, can be divided and merged again by means of calculation blocks, such as addition or multiplication (set of rules). Block diagrams offer a way of clearly displaying the interaction between various system components (such as controller and measuring element) and, to a limited extent, the signal flow through these components. However, time sequences in particular are displayed only in a rudimentary way, if at all. Even dynamic control flows between the components are difficult to see.

Models are used in the development of embedded systems at all stages of development, from system specification to verification and documentation. They also serve as a basis for communication between the individual disciplines, such as hardware development, software development, business analytics and system integration. Since different aspects of the system are considered in each development stage and each discipline, different modeling techniques have emerged that approach the task with different focal points—such as signal flow, system design or component dependencies, but also production process and cost structure. A basic understanding of working with models is therefore essential for the developer. Some modeling languages which are of particular importance in the area of embedded systems are introduced briefly below. For more detailed information, please refer to the corresponding literature.

An important distinguishing feature between models is the level of abstraction. For example, a model at system level describes in an abstract way what a system should generally do and which requirements it must meet. The interaction of the subsystems and how they implement this task is shown at module or group level. At the program and/or logic level, concrete implementation is described e.g. in a programming language or using hardware gates (see Fig. 12.1).

In addition to the abstraction level, models can also be classified according to their *view* of a system. A distinction is made between *structural models*, which rep-

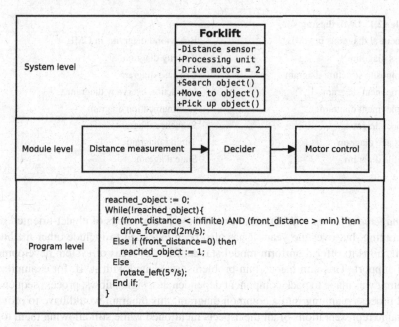

Fig. 12.1 Example of different abstraction levels

resent the structure and interconnection of components of the system, and *behavioral models*, which represent the processes within the system.

The following sections illustrate this using the autonomous forklift truck as an example. From a behavioral point of view, it should turn counterclockwise until the front distance sensor detects an obstacle and then drive up to it until the obstacle is directly in front of it. It should evaluate and record this as necessary.

From a structural point of view, this is a system consisting of a decision maker or a processing unit, a distance sensor and two drive motors that exchange information in that order and interact with an external obstacle.

Since, in the course of development, any model type can emerge, the result is often a heterogeneous overall model of the system created by merging these partial models, which supports the documentation substantially.

12.2   UML

In the different phases of development and maintenance of embedded systems, various models are needed for the most important aspects. In order to achieve coherence or comparability, a uniform, underlying structure of these models is helpful.

Unified Modeling Language (*UML*), version 2.2 which is considered here, offers an approach to such a structure. This modeling language, maintained by the *Object*

Table 12.1 UML diagrams

Structural diagrams in UML	Behavioral diagrams in UML
Class diagram	Activity diagram
Composite structure diagram	Use case diagram
Component diagram	Interaction overview diagram
Deployment diagram	Communication diagram
Object diagram	Sequence diagram
Package diagram	Timing diagram
Profile diagram	State diagram

Management Group (*OMG*), is originally based on methods of object-oriented programming, but over the years it has also been extended to include other methods. UML aims to offer a uniform model standard for all areas of system development and support. This aim has certain problems associated with it. If, for example, an attempt was made to pack component dependencies, signal flows, process sequences and process planning into a common diagram, this diagram would have to provide suitable representations of all the aspects mentioned while still allowing them to be classified clearly. This would lead to a multitude of modeling elements and a very confusing diagram. UML avoids this by offering a common "language basis"—the so-called metamodel—for currently fourteen special diagram types, instead of a uniform diagram of all aspects. These can be classified according to the above views, as shown in Table 12.1.

Some of these diagrams are examined in more detail below as examples. Further diagrams and a deeper insight into UML can be found e.g. in [1] or [2].

12.2.1   Activity Diagram

Activity diagrams are used to describe processes within a system. They can be used to display both the control flow and the signal/object flow. They use elements of Petri nets (see Sect. 12.4). Control and object tokens are passed through the diagram. Only the actions that have a corresponding token are executed. This makes it possible to represent concepts such as concurrency and asynchronous communication mechanisms. Activity diagrams usually consist of the following components:

- **Activity** The behavior to be displayed that contains the individual steps. It is represented by a rectangle with rounded corners, with the identifier in the upper left corner.
- **Action** An (elementary) step in the execution of the activity, represented by a rectangle with rounded corners.
- **Decision node** A branch in the activity flow that is linked to a condition, represented by a diamond.

- **Union nodes** Offer the possibility to merge several flows into one. They are also displayed as diamonds.
- **Parallelization and synchronization nodes** Used to represent concurrency. Thus one control flow becomes several or vice versa. In contrast to the union node, all incoming flows must be fulfilled in order for the outgoing flow to be fulfilled too. A bold line is used for representation (see transitions in Petri nets).
- **Objects** are the input and output interfaces of the activity. They are displayed as rectangles that extend beyond the edge of the activity.
- **Input and output pins** Represent the output or acceptance of object tokens. They appear as squares on the sides of actions.
- **Control and object flow** Displays the activity history in the form of arrows. Object flows correspond to data flows in progress diagrams, while control flows represent the transport of the control tokens through the activity. Graphically, they differ in that an object flow takes place between two pins or contains an object, whereas a control flow is directly linked to the actions.
- **Start and end nodes** Display the entry or end point of the activity. The start node is represented by a black circle, the end node by a bordered black circle (see places in Petri nets).

This is illustrated by the example shown in Fig. 12.2. From the start node, the virtual control node passes through a join node and reaches the action "Check distance". This receives an object token from the input object "Distance sensor values". The object flow is shown in gray here to highlight it. Strictly speaking, the input pin at "Check distance" is superfluous, since the flow already contains an object ("Distance sensor values"), but it can be added for clarification.

When this action is complete, the virtual control token reaches a decision node. Here it is decided whether or not the distance is greater than the measuring range; if so, it means that no object was detected. If no object is detected, the action "Turn counter-clockwise" is performed, i.e. the control token is passed to it. After its completion, it passes the control token through the union node again to "Check distance" to initiate a new check.

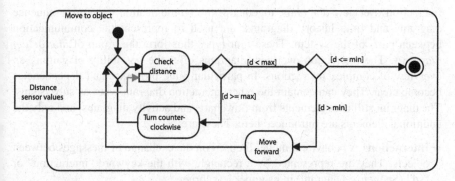

Fig. 12.2 Example of an activity diagram for searching an object and driving towards it

If an object is detected this time, the next decision node checks the distance to determine whether it has already been reached. If this is not yet fulfilled, the control token is passed on to the action "Move forward", which also initiates a new check of the distance after its completion. If the measured distance is minimal, it is assumed that the object is directly in front of it, and the control token is passed to the end node and the activity is thus completed.

12.2.2 Timing Diagram

Timing diagrams are used to visualize states over time. They correspond to the typical representation of an oscilloscope. The time is plotted on the x-axis, while the state of the objects to be examined is plotted on the y-axis. Several objects can be applied on top of each other. Correlations between temporal changes are indicated by arrows. Diagram elements include:

- **Objects/Elements** Correspond to objects or entities from object-oriented programming. Their possible states are of particular interest for the timing diagram. They are generally specified by the notation *name:type* and separated in the timing diagram by a continuous horizontal line. If only one element of a type occurs, the name is often omitted.

Here, too, the search for and approach to an object by the forklift truck serves as an example. In Fig. 12.3, the temporal relationship between state transitions during the distance check and the corresponding control commands to the engine control is represented. The visible delay can be caused, for example, by signal propagation times, intermediate calculations or similar.

12.2.3 Interaction (Overview) Diagrams

Interaction overview diagrams, together with communication diagrams, sequence diagrams and time history diagrams, are used to represent the communication between parts of the system. These four types thus form the group of *interaction diagrams*. The interaction overview diagrams offer the possibility of gaining an overview of complex interactions. In particular, time processes and dependencies become clear. They can contain the other interaction diagram types as subsystems. The diagram still uses elements from flowcharts and activity diagrams, and only the additional elements are introduced here. These are:

- **Interactions** A behavioral model focused on the exchange of messages between objects. They are represented by a rectangle with the keyword "interaction" or "sd" (Sequence Diagram) in its upper left corner.

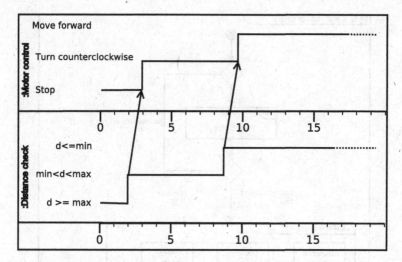

Fig. 12.3 Example of a timing diagram for searching an object and driving towards it

- **Interaction references** References to an interaction that has been defined else-
  where so that multiple interactions can be displayed clearly in a common diagram.
  They are displayed as a rectangle with the keyword "ref" in its upper left corner.

In Fig. 12.4 the above example is extended. It shows the interaction between the
three sub-interactions "Find object", "Move to object" and "Pick up object". In
order to avoid making this diagram unnecessarily complicated, the first and last sub-
interactions are only specified as references and only "Move to object" is explicitly
displayed as a sequence diagram.

12.2.4 Class Diagram

Class diagrams are an important tool in object-oriented development. They visualize
the relationship of *classes*, that represent an abstract model of a series of similar
objects. This static structure usually corresponds to the structure of the actual code
of an object-oriented program. This is less about the behavior of the system than
about its structure. Class diagrams consist of the following elements:

- **Classes** These are the basic objects of the diagram. They are displayed as rectangles. Each class has a name. In addition, it can contain *attributes*, that is, properties
  or parameters, which are shown under the name and separated from it by a slash.
  Furthermore, it can have *operations*, i.e. typical actions that the class can execute,
  which in turn are found below the attributes. Both attributes and operations can
  have different *visibilities*, which are represented by preceding operators (e.g. +
  for public, − for private, # for protected).

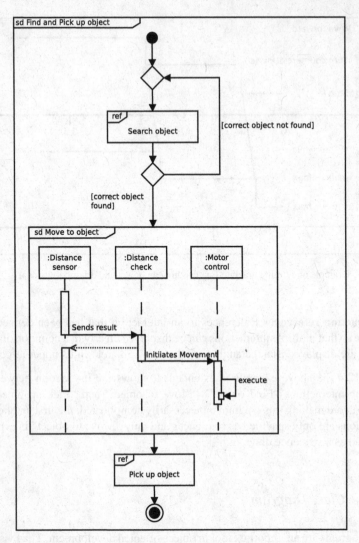

Fig. 12.4 Example of an interaction overview diagram for searching an object, driving towards it and picking it up

- **Relationships** These describe how the classes relate to each other. There are five types of relationships: *Dependency, Aggregation, Assoziation, Composition* and *Inheritance/Generalization*. They are distinguished by different arrow types. The *dependency* shows that one class uses another (dashed line with filled arrow end). *Aggregation* and *composition* show that a class consists of, or contains, others. The difference is that the classes contained in a composition cannot exist without the composite class (there can be no room without a building, but the building consists of rooms). Both are drawn as a solid line with a diamond-shaped arrowhead,

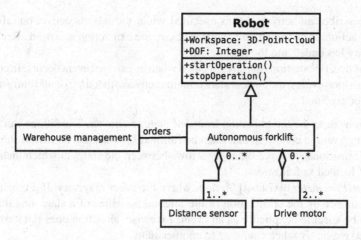

Fig. 12.5 Example of a class diagram for the relation regarding the autonomous forklift truck

which is solid for composition but not for aggregation. The *assoziation* represents a general relationship between two classes (represented as a solid line, undirected). Finally, *inheritance* is an important concept of object-oriented programming. It expresses the fact that a class is a special form of an original class and that it shares its properties (attributes and operations) (solid line with an unfilled arrowhead).

An example is the autonomous forklift truck, as shown in Fig. 12.5. The class "Autonomous forklift" is a specialization of (inherited from) the class "Robot", so it has the attributes "Workspace" and "DOF", and the operations "startOperation" and "stopOperation". It is associated with the class "Warehouse management" and is an aggregation of "Distance sensors" and "Drive motors", among others.

12.2.5 State Diagram

The state diagram is the UML-standardized representational form of (deterministic) finite automata. It combines other forms of presentation and extends them to include further options, such as hierarchically nested states. In contrast to other machine diagrams, even complex machines can therefore be displayed clearly. Important elements of the state diagram are states and transitions.

The state diagram represents a hypothetical machine (the automaton), which can have only one *state* or only a defined, finite set of system states at any time. They are displayed in the diagram as rectangles with rounded corners. States can have internal activities, which appear under the name of the state with the following keywords:

- **Entry** describes an action that is executed as soon as the state becomes active.
- **Exit** describes an action that becomes active as soon as the state is left.

- **Do** describes an activity that is executed while the state is active, but after the entry action. If no transition conditions exist, the exit action is started after the do activity has ended and the state is left.
- **Event** describes actions that are executed when an external event occurs. In contrast to transition events, the current state remains active so that the exit and entry actions are not executed.

Transitions describe the transition from one state to another. Transitions can have a triggering *event*, a condition that must be fulfilled *(guard)*, and an *action* started by them. Transitions are represented by arrows between the states at which their event chain is located as follows:

Event(Parameter) [Guard]/Action, where the order may vary. If a transition is triggered only by the termination of the internal activities of a state, this does not have to be specified explicitly as an event. Likewise, an action does not have to be specified explicitly when changing to another state.

In Fig. 12.6 you can see a state machine of this sort for the problem described above. The robot is initially in the state "Robot stands still". A green LED is activated when this state is reached. Now the robot checks the value measured by the distance sensor. This internal activity could be interrupted by an internal or external event, but this does not happen here.

At the end of the check, one of the transitions to one of the other states is executed, depending on the value of the distance "d". On transition to the state "Robot turns counterclockwise", the condition "Check complete" and the resulting action "Start countercl. rotation" are indicated in brackets again, which otherwise are implicitly derived from the internal action of the states and their designation.

It is important that there is now a transition for all values that "d" can get, since otherwise the machine will be stuck in its current state after checking the distance. On leaving the respective state, the appropriate LED is switched off. In the case of transitions referring to the initial state (e.g. if the forklift truck is in the "Robot turns counterclockwise" state and the transition is triggered with the condition "d > max(d)"), this means that the respective LED is first switched off and then on again for the duration of the transition (here assumed to be 0). If the forklift truck is now

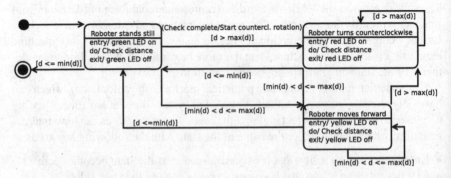

Fig. 12.6 Example of a state diagram for searching an object and driving towards it

in the "Robot stands still" and determines that the distance "d" is less than or equal to its minimum, the process is completed.

12.2.6 Component Diagram

The component diagram is a structural diagram. It can be used to display the relationships of components within the system. The components are represented as individual blocks, if necessary with additional information about which classes they represent and which interfaces they provide. The components are interconnected by different arrows which can represent communication paths, dependencies or inheritance (generalization). This makes them similar to block diagrams. Important elements include:

- **Components** These are abstract elements of the system, comparable to classes. They are represented as rectangles and contain the keyword $<<$ *component* $>>$. They can be displayed as black boxes, or their contents can be shown in the form of further diagrams.
- **Interfaces** Show the communication between components. A distinction is made between *available* interfaces (empty circle at the end of a line) and *required* interfaces (open semicircle at the end of the line). Interfaces can also be delegated to a higher-level component. This is represented by an arrow with a dotted line running from an interface to a port of the parent component.
- **Artifacts and realizations** Artifacts are resources that a component uses to fulfill its purpose, realizations represent the concrete implementation of a component or a part of a component. They are identified by the keywords $<<$ *artifact* $>>$ and $<<$ *realization* $>>$.

In Fig. 12.7, the internal structure of the forklift truck with the essential components for the task is shown as a component diagram. As an interface, the "distance sensor" offers the measured values which the "Processing unit" receives. An "Industrial PC" can be used here as a realization of the processing unit. After evaluating the values, the processing unit offers speed values for the two motors. The interfaces leaving the system are now delegated to the higher-level component. The internal components are shown here as a black box; their internal structure is not visible.

12.3 SDL

Specification and Description Language (*SDL*) is a modeling language used in telecommunications. Although originally designed to handle large distributed communications networks, it has evolved over the years into a multifunctional system description language. A comprehensive introduction to this can be found, for example, in [3]. Compared to UML, SDL is much more formal. This is also expressed

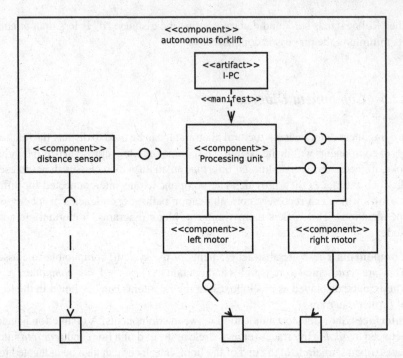

Fig. 12.7 Example of a component diagram for the structure of the autonomous forklift truck

in the fact that an equivalent textual representation is provided alongside the graphical representation, so that SDL models can exist both as diagrams and in written form. Either representation can be transferred into the other. This has the advantage that SDL models can be simulated and verified directly and can also be converted into executable code automatically (e.g. C/C++, or VHDL). This virtually provides a self-documenting implementation. On the other hand, some of the flexibility of UML is lost and some views of the system are not available. The SDL model is also less intuitive than UML. However, there are now tools that can be used to convert (a subset of) UML to SDL and vice versa. SDL's underlying model assumes that the system consists of a number of extended state machines that run in parallel and independently of each other and communicate with each other using discrete signals. This foundation enables efficient modeling of real-time systems, which makes SDL particularly useful for embedded systems. SDL models are divided into four levels of abstraction: the system level, the block level, the process level and the procedure level. The modeled system is represented at system level as interconnected, abstract blocks. This makes it possible to ignore implementation details that are irrelevant for the system structure and to maintain an overview of the overall structure. The individual blocks contain a number of processes at block level, which can be regarded as embedded hierarchical state machines at process level. These machines, in turn, can consist of sub-machines that represent the above mentioned procedures, which ultimately represent local, global or cross-process functions in the network. SDL also distinguishes between the static structure, which consists of the interconnection of blocks using appropriate channels, and the behavioral structure, which represents the interaction of processes using signals and their internal structure. However, these are used in a heterogeneous overall model. A special feature is that there is no implicit global memory in SDL. Each process has its own address space. This reduces dependencies within the architecture and makes the system more robust.

Communication between the system components is realized by means of so-called *gates*, which are connected via *channels*. Channels are used for the transmission of signals between the components, which are usually subject to a communication delay. This makes signal transfers asynchronous, which allows a very realistic representation of the communication over time. However, this forces incoming signals in a component to go first into a FIFO queue and they must be processed one after the other. They serve as triggers for state transitions in the state machines included. Important elements include:

- **Declarations** These serve to declare signals, variables and timers in the concrete (and underlying) model (rectangle with angled corner).
- **Input and Output** Indicate that at this location an incoming signal is awaited or a signal is being transmitted (rectangle with triangular notch [input] or bulge [output]).
- **Processes** These are the executing program units themselves (rectangles with corners cut off, as a reference there may be several inside one another, see Fig. 12.9).
- **States** These correspond to the states of an automaton (rectangle with convex sides).
- **Process calls** Here another process/procedure is opened (rectangle with double horizontal lines).

The Figs. 12.8, 12.9, 12.10 and 12.11 show the example system on various levels.

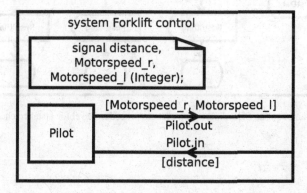

Fig. 12.8 Example of an SDL diagram on system level for the control of the autonomous forklift truck

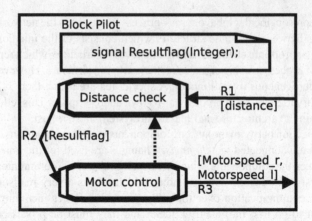

Fig. 12.9 Example of an SDL diagram on block level for the control of the autonomous forklift truck

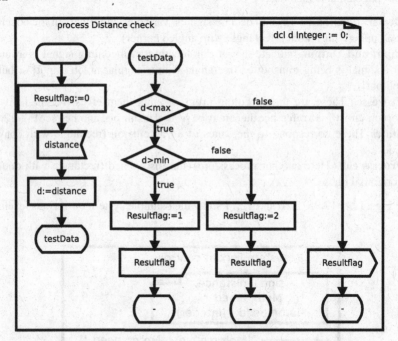

Fig. 12.10 Example of an SDL diagram on process/procedure level for distance checking

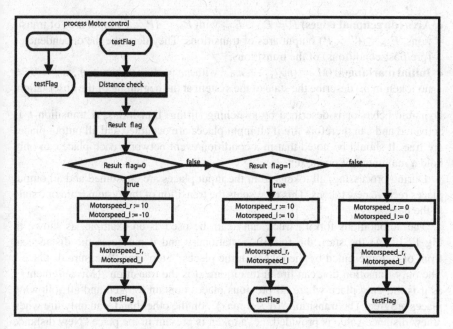

Fig. 12.11 Example of an SDL diagram on process level for the motor control

12.4 Petri Nets

In 1962 C. A. Petri developed a graphical description tool in automata theory, the Petri net. Today it is one of the most important descriptors for concurrent systems (software and hardware) and it has been extended several times (to include, for example, stochastic and time-extended networks). Petri nets have been extensively investigated in terms of their theory (and can easily be converted into mathematical models for analysis/simulation) and serve as qualitative statements about system behavior (resource conflicts, freedom from deadlock, time aspects, ...).

Over the years, various versions and extensions of the classic Petri net have been developed. The *condition/event net* is explained here in more detail, as one of the most used cases. More information can be found e.g. in [4]. A *Petri net* (condition/event net) is a bipartite directional graph to which the following applies: $PN = (P, T, E, M_0)$. It consists of

- **Places** ($P = \{p_1, \ldots, p_n\}$) model predicates or resource types, or possible system states that may contain *tokens*.
- **Tokens** are marks, that indicate whether a predicate is fulfilled or a resource is present.
- **Transitions** ($T = \{t_1, \ldots, t_m\}$) model active elements (e.g. processes).

- **Arcs (directional edges)** $E = E_{in} \cup E_{out}$ with $E_{in} \subset (P \times T)$ input arcs of transitions, $E_{out} \subset (T \times P)$ output arcs of transitions. They describe the dependencies (pre/post-conditions) of the transitions.
- **Initial marking(s)** $(M = \{m_{01}, \ldots, m_{0n}\}$ with $m_{0i} = 1 \Rightarrow$ token in p_i, $m_{0i} = 0 \Rightarrow$ no token in p_i) describe the state of the system at the beginning of the observation.

Dynamic behavior is described by switching (firing) transitions. A transition t_i is activated and can therefore fire if all input places are occupied and all output places are free. It should be noted that in a condition/event network, each place can only hold a maximum of one token.

During processing, all tokens from the input places are consumed and all output places receive new tokens. This represents the transition of the system from one state to the next.

The autonomous forklift truck can again be used as an example, as shown in Fig. 12.12. At the start, the forklift is stationary and it measures the distance in front of it, represented by the tokens in the places "Stop" and "Measure distance". The only transition that can fire at this moment is the transition "(Measurement)", as it is the only place where all previous places contain a token and all following places are free. The transition "(d >= max)", on the other hand, can only fire when a new distance value is provided, i.e. a token is present in the place "New distance value". Note that firing of the transitions depends only on these tokens, the conditions included in the labels serve only as a guide, but are not explicitly modeled in the Petri net. This results in a recurring change between the three states "Stop", "Turn counterclockwise" and "Move forward", with a wait between all the changes for a new distance value to be provided. For reasons of clarity, the possible transitions from "Turn counterclockwise" to "Stop" or from "Stop" to "Move forward" have not been modeled here.

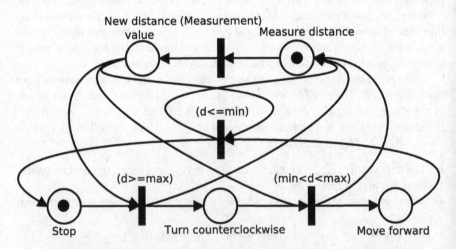

Fig. 12.12 Petri net of the autonomous forklift truck example without time extension

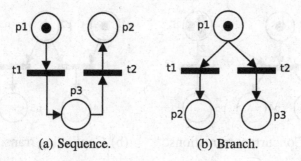

(a) Sequence. (b) Branch.

Fig. 12.13 Non-determinism in Petri net transitions

Petri nets are not always deterministic. On the one hand, without extended time representation (e.g. in time-extended Petri nets) no conclusions can be drawn about the time at which a transition fires. On the other hand, at an input place for two transitions only one of them can fire and no explicit model is created which does this. In Fig. 12.13, for example, the sequence can be seen as a deterministic sequence of transitions and as a non-deterministic set of branches, since it is not clear which transition fires without further conditions.

Condition/event systems are equivalent to finite automata. There are a large number of extensions, e.g. place/transition networks (with negation), predicate/transition networks as well as time-based and stochastic networks.

Place/transition nets are of particular interest here. They are condition event networks extended to include multi-token places (merging of several places), weighted edges (merging of several edges) and (possibly) input edges with negation (inhibitor edges).

This changes the rules for firing transitions. A transition t_i is activated in a place/transition net if *(a)* all places p_j of the (not negated) input arcs (p_j, t_i) (with weighting w_j) contain at least w_j tokens, or *(b)* places p_k with inhibitor input arcs (p_k, t_i) (with weighting w_k) do not contain any tokens (or contain less than w_k tokens).

In the literature, there is also the restriction that by firing a transition a maximum of tokens per place may not be exceeded. The place transition nets are still non-deterministic, but equivalent to the Turing machine. This is important when Petri nets have to be analyzed, because different types of nets can handle the same system states differently.

12.4.1 Analysis of Petri Nets

As mentioned above, Petri nets are very well suited for formal analysis of certain system properties. Conclusions are drawn from network properties about system properties. These can be structural properties (static properties) such as conflicts, or marking-related properties (dynamic properties) such as:

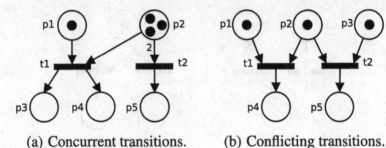

(a) Concurrent transitions.   (b) Conflicting transitions.

Fig. 12.14 Concurrency and conflict in Petri nets

- **Reachability** A marking M_n is reachable from M_0 if and only if $n = 0$ or there is a switching sequence $s = (t_1, t_2, \ldots, t_n)$ that converts M_0 to M_n. A marking may be desired or undesirable.
- **Liveness** A PN is alive if for each marking M_n, that is reachable from M_0, all transitions of the net can be activated by a marking M_k that is reachable from M_n. Liveness prevents deadlocks.
- **Boundedness** A place is bounded if the number of tokens it contains does not exceed an upper limit N. Places can be realized by buffers of the size N.
- **Persistence** A PN is persistent if for each pair of activated transitions no transition is deactivated by switching of the others.
- **Fairness** A pair of transitions is fair if there is an upper limit of executions of a transition before the other switches.
- **Conservatism** A PN is conservative if the number of tokens in the net is always constant. This is important when, for example, tokens model objects in closed systems.

The analysis can be used to investigate temporal processes in concurrent systems.

Critical points in the analysis usually occur when several transitions (could) fire. Specific designations have therefore been introduced for these cases. Transitions are *concurrent* if they are activated at the same time and can switch in any order without deactivating each other. It is therefore not relevant locally which transition fires first. Two transitions *conflict*, on the other hand, if both are activated but one is deactivated by switching the other. The place that is the cause of the conflict is called the decision point. Here the order of firing can affect further system behavior significantly, which is why these places require special attention and may need further specifications. Both cases are illustrated in Fig. 12.14.

Figure 12.15 shows the *confusion* between transitions. This particular type of conflict is characterized by the fact that the concurrency of transitions is affected by the switching of another transition.

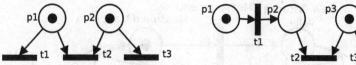

(a) Symmetric confusion.   (b) Asymmetric confusion.

Fig. 12.15 Confusion between transitions

12.4.2 Timed Petri Nets

With embedded systems, real-time criteria often play a major role (see Sect. 13), which is why the time sequences of the systems must also be considered. In order to investigate these, time-extended Petri nets were introduced. For more in-depth information, we recommend e.g. [5]. Places (e.g. duration of presence of tokens) and transitions (e.g. circuit symbols) are suitable as time attributes.

An execution time/delay c_i is assigned to a *timed* transition t_i (time from the start of switching until output stations are occupied). Alternatively, an interval $[c_{min}, c_{max}]$ can be assigned within which c_i must stay. For *timed* places, calculation processes that are started when a transition is switched and take c_i time units are assigned to the output places.

An example of this is the *Duration Petri net* (*DPN*). Each transition is assigned a switching duration D. This means that if the transition t_i fires at the time τ, the tokens from the previous places are removed, but they are not added to the subsequent places until the time $\tau + D(t_i)$. In the interim, these tokens exist virtually in the transition. To prevent the exploration space for the net analysis from becoming infinitely large, a virtual clock is assumed that counts in constant, discrete time steps. This means that the switching times can only be multiples of these discrete time steps.

Another change that occurs in many time-dependent Petri net models is the fire compulsion of transitions. While in a normal Petri net the transition fires at any time after its activation, in a DPN it must fire as soon as possible. This ensures that the significance of the assigned switching duration is maintained. Thus, processes that require a certain execution time can be modeled more effectively using DPNs, which is particularly helpful for concurrent paths. This is illustrated by the example of the forklift truck. In Fig. 12.16, the Petri net represented in Fig. 12.12 has been extended to include transition switching durations. The transition "(Measurement)" now requires three time steps to simulate the duration of the measurement, the other transitions each require one for the computing time. Thus at the beginning the token is removed from the place "Measure distance" and is not added until three steps later to the place "New distance value". Only now is "(d >= max)" activated, removes all tokens from "Stop" and "New distance value" and adds them one step later in "Turn counterclockwise" and "Measure distance".

This results in a defined time between the transitions of the driving status of the forklift truck of four time steps each. Although this does not represent the real system

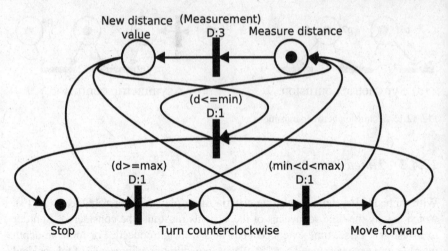

Fig. 12.16   Petri net of the autonomous forklift truck example with a time extension

behavior, in which the external conditions for the distance value play the main role, it does allow a temporal analysis of the system. For example, a maximum speed for the forklift truck can be derived without much effort, since there is a response to a different distance value at the earliest four time steps later.

References

1. Kossiakoff A, Sweet WN, Seymour SJ, Biemer SM (2011) Systems engineering principles and practice. Wiley, New Jersey
2. Seidl M, Kappel G, Scholz M, Huemer C (2015) UML classroom. Springer International Publishing
3. Ellsberger J, Hogrefe D, Sarma A (1997) SDL: formal object oriented language for communicating systems. Prentice Hall, Harlow [i.a.]
4. Chen WK (2004) The electrical engineering handbook. Elsevier, Burlington
5. Popova-Zeugmann L (2013) Time petri nets. Springer, Berlin

Chapter 13
Real-Time Systems

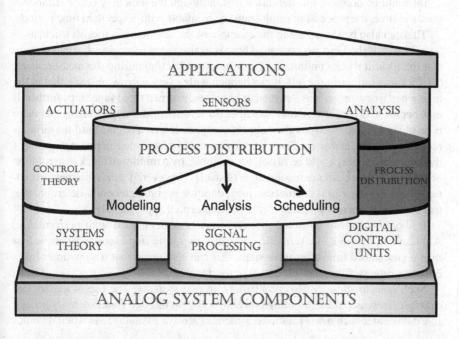

This chapter presents procedures that can be used to ensure the real-time capability of a system. The first step is the allocation process that determines how processes can be meaningfully divided among distributed resources. Then various scheduling procedures are covered with their respective strengths and weaknesses, and include planning by searching, earliest deadline first, least laxity and rate-monotonic scheduling. Thus it is possible to schedule periodic and sporadic processes as well as pre-emptible and non-preemptible processes in a meaningful order, provided that the available resources allow this.

© The Author(s), under exclusive license to Springer Nature Switzerland AG 2021   339
K. Berns et al., *Technical Foundations of Embedded Systems*, Lecture Notes in Electrical
Engineering 732, https://doi.org/10.1007/978-3-030-65157-2_13

13.1   Real Time in Embedded Systems

Many embedded systems must comply with real-time conditions. Failure to perform the calculations within a specified period of time may result in severe quality degradation (e.g. if audio or video quality suffers from dropouts or jumps), or may even result in physical damage to the user (e.g. if systems in cars, trains or aircraft do not behave as intended). Embedded systems the time behavior of which can lead to a disaster are called hard real-time (systems). All other violations of time conditions are called soft real-time systems (see Fig. 13.1).

Although outdated, DIN 44300 offers a thorough definition of real-time operation: "Real-time operation is the operation of a computer system in which programs for processing incoming data are constantly ready for operation in such a way that the processing results are available within a specified period of time. Depending on the application, the data can be distributed randomly or at predetermined times."

It should be noted in this definition that, although the data may occur at random points in time, the processing results must be available within specified time periods.

This can also be shown using the example of the autonomous forklift truck, as in Sect. 12.4.2. If the Petri net presented here is examined in more detail, it can be seen that the forklift truck continues to carry out its old action during the measurement. So if it turns further to the left, it can happen with a very narrow obstacle that at the time when a new target was detected, it has already turned three time steps further. It now begins to drive forward and only after three further time steps realizes that there is no obstacle at all. Now another, almost complete left turn begins until the target is recognized again, and so on. Since we have assumed a fixed measurement duration of three time steps, this could be offset, for example, by a minimum clockwise rotation of the forklift truck after detection of an obstacle. However, if a measurement exceptionally lasts longer or if the measurement duration is completely non-deterministic, this can become a real problem if it is not intercepted appropriately.

It should therefore be noted that real-time is not confused with performance. Although an *average measurement time* after two time steps sounds better at first than a guaranteed time after three steps, this can also mean that it sometimes takes only one time step, another time five, or even ten.

Such inaccuracies are also problematic in real systems. The use of caches, for example, increases the computing power of a system through faster memory access. In a data center with n workstations, a cache is always advantageous when it comes

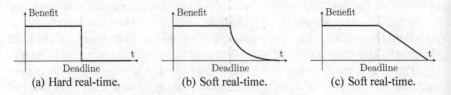

(a) Hard real-time.   (b) Soft real-time.   (c) Soft real-time.

Fig. 13.1 Illustration of hard and soft real-time

to increasing throughput. With an aircraft controller, on the other hand, deterministic timing of processes is necessary for scheduling. Since cache access times cannot be predicted for process changes, they cannot be used here. Similar problems can occur due to *direct memory access* or poorly integrated interrupts. Therefore, different procedures and algorithms have been developed to ensure adherence to time limits for the execution of several processes. Further information about this is provided, for example, by [1, 2].

Processes in real-time systems can be divided into clock-based processes (cyclical, periodic) and event-based processes (aperiodic, sporadic). The response time of the embedded system is therefore related to the "time scale" of the technical process. In clock-based processes, measured values are recorded at a fixed sampling rate, the execution time of the processes is determined by the sampling rate. In event-based processes, state changes of the system occur as a result of interrupts. The computer system must react to the interrupt within a specified time. Aperiodic behavior therefore results, since events occur non-deterministically.

Research and teaching usually consider the concept of hard real time only. The idealized rule here is, that there are no transgressions of time conditions under any circumstances. This idealized requirement is not always realizable, e.g. in exceptions (failure of a computer, network, ...).

For hard real-time systems the following should apply: $P(r + \Delta e \leq d | B) = 1$, with

- **r, d** Start time (*r*eady) and (*d*eadline) of a task,
- Δ**e** Execution time of a task,
- **B** Constraints.

This means, that the probability of the timely completion of a task is 1 if constraint B (no failures etc.) is met. Δe includes all predictable times (e.g. regular timer interruptions, disk access etc.).

For the rest of the chapter, it should be noted that no distinction is made between task, process and thread, since these have different connotations depending on the situation/operating system/intention and the corresponding literature remains ambiguous in this respect. Since purely abstract planning algorithms are described in any case, we will speak consistently about *processes* below.

13.2   Allocation

The problem of process scheduling cannot be considered completely separate from that of allocation, which is therefore at least addressed here. During the execution of (software) programs in embedded systems, they often use various resources. These can be memory, GPUs or parallel processors, for example, but "independent" computing resources such as hardware multipliers, DMA, multiple caches etc. are also increasingly being used within the processing units. The increased parallelization of program execution in this way is expected to result in higher data throughputs, which in turn will result in faster processing of the task on which the program is based. This can become problematic if, in the course of this paradigm, the program is also divided into several parallel processes, since these may now compete for the specified resources. In addition, "administration" of parallel resources and their allocation requires additional effort. This is the task of resource allocation. Various objectives can be pursued, such as:

- utilizing the available resources (e.g. several processors) as evenly as possible,
- optimal use of individual resources,
- the fastest possible execution of processes with high priority,
- compliance with real-time conditions,
- minimization of allocation costs,
- guaranteed avoidance of deadlocks,
- ...

The fullest possible use of existing resources (in compliance with the performance requirements) is regarded as the objective of allocation below, since this usually achieves most of the specified objectives, taking account of the scheduling algorithms below. All resources are regarded as "processors", but the approaches presented can also be used in other allocation areas, such as mainframes or microcontrollers. Graph-based solution methods have proven to be helpful in providing a meaningful overview of the various allocation objectives.

13.2.1   Allocation as a Graph Problem

The objective of allocation can be understood as a cost optimization problem. The costs consist of the execution times of the processes and the switches from one process to the next as well as to another processor. It is assumed that the communication or context switch of processes on the same processor is negligible compared to the corresponding costs when changing processor. This means the task is to minimize the communication costs between processors. For this purpose, flow graphs have proven useful as modeling tools. A flow graph is created with:

- **Terminal node** processors P_i (e.g. 2 processors as source and sink as in Fig. 13.2),
- **Inner node** processes T_i,
- **Edges** directed, ET (connection graph of processes, continuous lines in Fig. 13.2) and EP (edges from processors to all processes, black dashed or dotted lines in Fig. 13.2),
- **Weights** ET (communication costs) and EP (costs for *not* executing the specified process on the processor -> inverse costs),
- **Cost calculation** Example (inverse costs w):

$$w(T_i, P_k) = \frac{\sum_{j=1}^{n} E_{P_j}^{T_i}}{n-1} - E_{P_k}^{T_i}$$

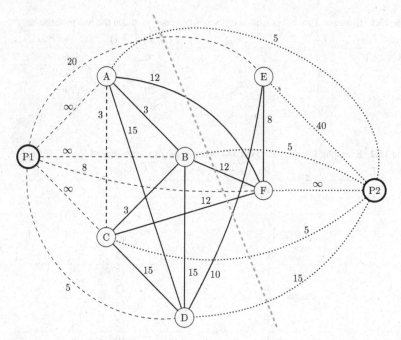

Fig. 13.2 Allocation flow graph for the Tables 13.1 and 13.2

with $E_{P_j}^{T_i}$: Execution costs of T_i on P_j.

Allocation here is the calculation of a special "allocation cut-set". The allocation cut-set is the subset of all edges whose intersection divides the graph into n disjoint subgraphs, where each subgraph contains a terminal node (= processor) (e.g. areas separated by a grey dashed line in Fig. 13.2). To optimize the costs, the system searches for the cut-set with the lowest total costs (sum of the costs/weights of the outgoing edges), the so-called *min-cut*. This therefore represents the allocation of processes to the corresponding processors that incurs the lowest combined execution and communication costs. If we look only at the times, it is the one that is executed the fastest.

For clarification, the forklift truck can be considered again. Let's assume that in a certain situation it should follow a trajectory based on received GPS coordinates and react to obstacles by means of three infrared sensors. A microcontroller, to which the infrared sensors are directly connected, and an industrial PC, which in turn is directly coupled to the power electronics of the motors, are available as processing units. The GPS module is connected to both processing units via a bus. Furthermore, the following processes come about:

- **A–C** Query of the measured values of the infrared sensor 1–3,
- **D** Query of the GPS position,
- **E** Calculation of the current position difference from the trajectory,

Table 13.1 Example: Process execution costs and "inverse" costs on the two processors

Process T_i	$E_{P_1}^{T_i}$	$E_{P_2}^{T_i}$	$w(T_i, P1)$	$w(T_i, P2)$
A–C	5	∞	∞	5
D	15	5	5	15
E	40	20	20	40
F	∞	8	8	∞

Table 13.2 Example: Communication costs between the processes T_i and T_j

Process T_i	Process T_j	$w(T_i, T_j)$
A	B	3
A	C	3
A	D	15
A	F	12
B	C	3
B	D	15
B	F	12
C	D	15
C	F	12
D	E	10
E	F	8

Table 13.3 Example: The three possible cut-sets

Processes on P_1	Processes on P_2	Total cost
A, B, C	D, E, F	115
A, B, C, D	E, F	**104**
A, B, C, D, E	F	122

- **F** Activation of the motors.

In addition, the execution and communication times of these processes on the two processors specified in Tables 13.1 and 13.2 result, whereby the microcontroller is designated by P_1 and the i-PC by P_2.

The resulting graph can be seen in Fig. 13.2. While it is quite obvious that the infrared sensors have to be interrogated by the microcontroller (processes $A - C \rightarrow P_1$) and the motors have to be controlled by the i-PC (process $F \rightarrow P_2$), the distribution of the processes D and E is not clear. A comparison of the cut-sets results in the total costs shown in Table 13.3. As a result, despite the significantly higher execution costs of the process D on the microcontroller, it is still better to execute it there than on the i-PC because of the high communication costs in that case. This is shown in Fig. 13.2 as a gray dotted intersection line.

13.3 Processes and Plans

In order to be able to guarantee the real-time behavior of a system, the chronological sequence of processes and, if necessary, their allocation to the available resources must be planned. Real-time planning consists of the planning feasibility check, plan creation and plan execution phases.

- **Feasibility check** This checks whether a meaningful plan can be created at all with the known information about the processes involved and the available resources. A distinction is made between static and dynamic checks. In a static check, before the system starts a check is carried out to determine whether all the deadlines can be met. With the dynamic check, on the other hand, the deadline check is carried out during runtime. This is often necessary (e.g. due to new processes), but it does not always provide meaningful plans because of insufficient data. This procedure is therefore only suitable for soft real-time requirements or requires rejection of new processes.
- **Scheduling or Schedule Construction** Here the planner receives all the information to assign the processors to the processes and creates either a finished plan, or a set of rules for a plan.
- **Plan execution or dispatching** This is where the actual allocation of processors to processes (by the so-called *dispatcher*) takes place.

Creation of the plans is called *scheduling*. There is a multitude of algorithms for planning, depending on the requirements of the system. Only a few important algorithms can be covered in this book. The following restrictions have therefore been applied:

- Only single processor systems and simple *SMPs* (*symmetric multiprocessor systems*) are considered.
- The goal of the planning is always adherence to all (hard) real-time conditions.
- Non-interruptible (non-preemptive) and interruptible (preemptive) processes are considered.
- Process switching/context switch is generally not taken into account during planning.

In order to identify general rules for planning, the following abbreviations are used below:

- **Type of process P** Realization of a task. Example: Query of a temperature sensor and data transfer.
- **Instance of a type of process P_i** The task (type of process) can be used more than once. Every instance has its own data and specific embedding. Example: Coupling of several temperature sensors.
- **jth execution of a process instance P_i^j** The process P_i can be restarted after completion. This feature is applied especially for periodic processes.
- **Ready Time r_i** From this point in time on, the processor may be assigned to P_i.
- **Starting Time s_i** At this point in time, the actual execution of P_i begins.

- **Execution Time** Δe_i This resembles the pure calculation time of P_i.
- **Deadline** d_i At this point the execution of P_i must be completed.
- **Completion Time** c_i At this time, execution of P_i' actually ends.

An important point here is the process execution time, since it is required by the real-time planning, but depends on various factors such as the available hardware (processor power, coprocessors, etc.), the input data of the respective process, the availability of the necessary resources and higher-priority processes or interruptions. This makes it difficult to determine the exact execution time, which is why an estimate, the worst-case or a statistically averaged execution time is usually used.

In addition to the process properties already mentioned, the regularity of the occurrence of processes and their (non-)preemptibility are particularly relevant. Therefore a distinction is made for scheduling tasks between *sporadic processes* and *periodic processes*, and between *preemptive/preemptible* and *non-preemptive* processes.

13.3.1   Sporadic Processes

Sporadic processes are started at irregular intervals (aperiodic). In principle, they can occur at any time, which makes planning more difficult. An example of this are external interrupts, which can occur independently of the current system time. These can be triggered, for example, by pressing a button or by changing a sensor value. Therefore

- for non-preemptive processes: $s_i + \Delta e_i = c_i$
- for preemptive processes: $s_i + \Delta e_i \leq c_i$

Non-preemptive processes therefore start at the time s_i, are executed immediately and continuously for Δe_i and end at c_i (see Fig. 13.3). Preemptive processes, on the other hand, can be interrupted during the runtime—as the name suggests—and resumed at a later point in time, which is why the actual completion time c_i can be considerably delayed compared to continuous execution. Here Δe_i is scattered over $[s_i, c_i]$.

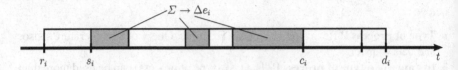

Fig. 13.3 Illustration of the times of sporadic processes

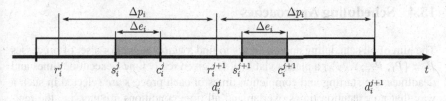

Fig. 13.4 Scheduling: Periodic processes

13.3.2 Periodic Processes

Figure 13.4 shows periodic processes. They are characterized by the fact that they recur at regular intervals. An example of this would be a polling algorithm that repeatedly checks the input buffer of a sensor. Periodic processes are easy to schedule because of these known repetitions.

$$r_i^j = (j - 1) \cdot \Delta p_i + r_i^1, j \geq 1 \tag{13.1}$$
$$d_i^j = j \Delta p_i + r_i^1, j \geq 1$$
$$= r_i^{j+1} \tag{13.2}$$

All the ready times and deadlines therefore depend on the first ready time r_i^1. However, the transition between sporadic and periodic processes is fuzzy for several reasons:

- all periods do not always start at the same time,
- the period duration can fluctuate (due to fluctuations in the technical process)
- deadlines are not identical to the period ends.

At worst, periodic processes must be treated as single sporadic processes.

13.3.3 Problem Classes

In order to decide which planning algorithm is best to use, certain boundary conditions must be taken into account. In addition to the interruptibility and periodicity of the processes, these are mainly the time at which the key figures for the processes are known and whether a complete plan or only a set of rules for such a plan is provided. Accordingly, the problem can be divided into the following classes:

- **Static scheduling** Data for scheduling previously known.
- **Dynamic scheduling** Data for scheduling is not available until runtime.
- **Explicit scheduling** Scheduler sends finished plan to dispatcher.
- **Implicit scheduling** Scheduler transfers only planning rules to dispatcher (e.g. priorities).

13.4   Scheduling Approaches

The aim of all scheduling approaches is to find a *viable plan*. For a set of processes $P = \{P_1, P_2, \ldots, P_n\}$ a plan is viable if, for given ready times, execution times and deadlines, the starting and completion times of each process are selected in such a way that no execution times overlap and all time conditions are met, i.e. the real-time requirements of all processes are met. Also the processor utilization u has to be considered, as not all approaches are usable under all load conditions.

$$u = \sum_{i \in P} \frac{\Delta e_i}{\Delta p_i}$$

The mere existence of such a plan does not provide any information about its calculation. Conversely, not every planning method always provides a useful plan, even if a plan theoretically exists. The concept of the *complete planning* was therefore introduced to be more precise.

- A static scheduling approach is called a complete scheduling approach if it delivers a viable plan, if one exists, for any processes $P = \{P_1, P_2, \ldots, P_n\}$.
- A dynamic scheduling approach is called a complete scheduling approach, if it delivers a usable schedule for any processes $P = \{P_1, P_2, \ldots, P_n\}$, if a static scheduling procedure with knowledge of all input data would have delivered a usable plan.

Table 13.4 provides an overview of the scheduling methods presented below.

Table 13.4 Overview of the scheduling methods presented below

Scheduling-strategy	Non-preemptive processes	Preemptive processes
Plan by searching	Complete, single processor, static, periodic and aperiodic	–
Earliest deadline first	Complete, single- and multi-core, static and dynamic, periodic and aperiodic	Complete, single- and multi-core, static and dynamic, periodic and aperiodic
Least laxity	Not complete, multi-core, static and dynamic, periodic and aperiodic	Not complete, multi processor, static and dynamic, periodic and aperiodic
Rate-monotonic scheduling	–	Complete on single processor, as long as $u < n\sqrt[n]{2} - 1$, static, periodic (aperiodic using the server principle)

13.4.1 Plan by Searching

Plan by searching is a static planning procedure, so it can only be used if all the required information is available at the time the plan is created. It is the most intuitive way to find a viable plan. There are no strategies or heuristics. The solution space is searched and there are basic planning considerations.

- Given: Set P of non-preemptive processes, $|P| = n$, with associated input data, i.e. ready times, execution times and deadlines
- Search for: Output of a sequence PL of tuples (i, s_i) with process index i, starting time s_i of P_i

Now all combinations in which the existing processes can be placed consecutively are simply determined and those that are unusable are excluded. The procedure runs as follows:

- Without considering ready times and deadlines, any sequence of processes is considered a correct plan. There are therefore $n!$ possible plans.
- A branch of the length k contains a set of processes $X_k \subseteq P$ with $|X_k| = k$ already scheduled. This creates a sub-plan PL_k for k processes.
- The following therefore applies to the scheduling of a process $k + 1$: $PL_{k+1} = extend(PL_k, i)$. A residual plan is generated, starting from the node PL_k.
- The newly created node PL_{k+1} thus contains the process P_i with $s_i =$ completion time of PL_k.
- In pseudocode the Algorithm 13.1 results.

Algorithm 13.1 Plan by searching without taking ready times or deadlines into account.

---

procedure SCHEDULE(PL_k, X_k)
  for all i in $P \setminus X_k$ **do**
    schedule (extendPL(PL_k, i), $X_k \cup i$)
                    ▷ Starting from PL_k, generate a residual plan including process k+1
  end for
end procedure

This can be seen clearly in the example in Fig. 13.5.

In order to take account of ready times and deadlines, each new subplan must be checked for compliance with the deadline. To this end, process P_i is scheduled as early as possible.

- Beginning with r_i, P_i is scheduled for the first gap into which it fits with Δe_i.
- If there is no such gap, P_i would miss its deadline d_i. Thus no plan with the previous decisions in the branch would be possible.
- The procedure is described in Algorithm 13.2, with

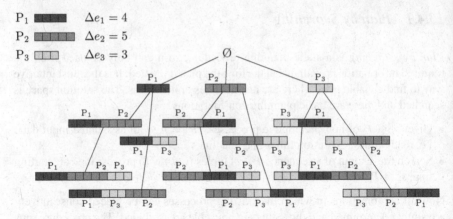

Fig. 13.5 Plan by searching without taking ready times or deadlines into account

feasiblePL(PL_k, i) tests, if P_i can be integrated into PL_k in a viable way at all, earliestPL(PL_k, i) schedules the process P_i starting from PL_k at the earliest possible gap.

Algorithm 13.2 Plan by searching taking account of ready times and deadlines.

---

procedure SCHEDULE(PL_k, X_k)
  for all (i in $P \setminus X_k$) AND feasiblePL(PL_k, i) **do**
    schedule (earliestPL(PL_k, i), $X_k \cup i$)
  end for
end procedure

This becomes clear in the example in Fig. 13.6, in which the processes from Fig. 13.5 have been extended to include ready times and deadlines. This not only causes gaps in the plan to appear (so-called *idle times*), but also makes various combinations unusable.

It can be shown that the procedure can be used to find a viable plan if such a plan exists. However, there are also viable plans that cannot be found with the procedure, in particular plans that require idle times between two processes are not found (see e.g. Fig. 13.7). Scheduling of non-preemptive processes with ready times and deadlines is thus complete, but in the area of complexity analysis (according to [2], pp. 161 ff.) it is also NP complete and therefore associated with high computing effort.

The procedure is not suitable for dynamic scheduling, as all ready times, execution times and deadlines must be known a priori. Although it is theoretically applicable to preemptive processes, in practice the solution space and therefore the calculation work would be disproportionately large, so that it cannot be described as suitable. Since the effort is also very high for static planning with non-preemptive processes, plan by searching does not play a major role in practice.

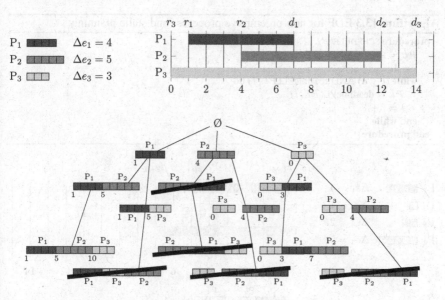

Fig. 13.6 Example for plan by searching

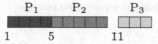

Fig. 13.7 Example for a viable plan, that can't be found by plan by searching

13.4.2 Earliest Deadline First

The *Earliest Deadline First* (*EDF*) method is widely used. This is mainly due to the fact that it is suitable for non-preemptive and preemptive processes as well as static and dynamic scheduling, i.e. it covers the essential problem classes. The procedure is still quite intuitive. The processor is always assigned to the ready process P_i with the shortest deadline d_i. If this condition applies to several processes, this leads to a random selection (usually solved with additional constraints/priorities). If, on the other hand, there are no processes ready, the processor remains idle.

For non-preemptive processes, the strategy is applied after each completion of a process, for preemptive processes before any possible process switch. This is made possible because the computational effort is limited to comparing the deadlines of all processes that are ready. The algorithms for preemptive and non-preemptive processes differ only slightly in the case of static planning. Under the assumption that the set of processes $P = P_1, \ldots, P_n$ is ordered by d_i, the result is Algorithm 13.3, where deadlinePL(PL, i) implements the above strategy. This is shown in the example in Fig. 13.8.

In the case of preemptive processes, the availability times and deadlines must now be taken into account in each time step. This increases the effort for the feasibility

---

Algorithm 13.3 EDF for non-preemptive processes and static planning.

procedure SCHEDULE(PL, P)
   $PL = \varnothing$
   $i = 1$
   while $(i \leq n)$ AND feasiblePL(PL, i) **do**
      $PL = $ deadlinePL(PL, i)
      $i = i + 1$
   end while
end procedure

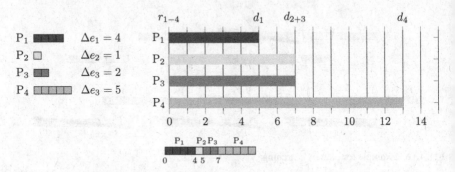

Fig. 13.8 EDF for static scheduling, non-preemptive processes

check (feasiblePL) in particular. For formal description, some calculation variables are introduced.

- Since the processes are preemptive, they are processed at intervals. These usually result from the cycle time of the system. The current interval of the process P_i is designated by m_i.
- The plan PL therefore represents a sequence of tuples $(i, s_{i,j}k, \Delta e_{i,j})$ with

$s_{i,j}$: Start of the interval j of P_i
$\Delta e_{i,j}$: Duration of the interval j of P_i.

- Thus the execution time Δe_i of the process P_i is divided into several parts $\Delta e_{i,j}$. Hence

$$\sum\nolimits_{j=1}^{m_i} \Delta e_{i,j} = \Delta e_i$$

- Furthermore, the process P_i should be completed before its deadline despite this division, and its first interval can begin at the earliest at its ready time. Hence:

$$r_i \leq s_i = s_{i,1} \leq s_{i,m_i} + \Delta e_{i,m_i} = c_i \leq d_i$$

- In order now to determine a plan for the next time step, all processes P_i must first be determined that are ready at the time t and their deadlines can still be met:

$$ready(t) = \{i \,|\, r_i \le t \le d_i \wedge rem(i, t) > 0\}$$

with

$$rem(i, t) = \Delta e_i - \sum_{j \in PL} \Delta e_{i,j}$$

- From this set ready(t), using $edf(ready(t))$ the process P_i with the shortest deadline d_i is now determined.
- It must now be taken into account that further processes with different ready times may have to be scheduled for this. The time of the next ready process is determined for this purpose:

$$nextAvail(t): \begin{cases} min\{r_i \,|\, r_i > t\} & \text{if}(\{r_i \,|\, r_i > t\} \ne \varnothing) \\ max(d_i) \text{ or } \infty & \text{else} \end{cases}$$

- In order to determine whether all relevant processes are scheduled, their remaining unscheduled execution time is considered:

$$AllinPL(t) \iff \left(\sum_{i \in P} rem(i, t) = 0 \right)$$

- This results in Algorithm 13.4 for preemptive processes and static planning, with $\neg feasible(i, t) \equiv t + rem(i, t) > d_i$.

Algorithm 13.4 EDF for preemptive processes and static planning.

---

procedure SCHEDULE(PL, P)
   $PL = \varnothing$
   $t = min\{r_i \,|\, r_i \in P\}$
   while $\neg AllinPL(t)$ **do**
      if $ready(t) = \varnothing$ **then**
         $t = nextAvail(t)$
      else
         $i = edf(ready(t))$
         if $\neg feasible(i, t)$ **then**
            BREAK
         end if
         $\Delta e = min(rem(i, t), nextAvail(t) - t)$
         $PL = PL + (i, t, \Delta e)$
         $t = t + \Delta e$
      end if
   end while
end procedure

This can be seen in the example in Fig. 13.9.

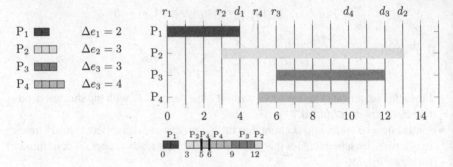

Fig. 13.9 Example for EDF with static scheduling of preemptive processes

In practice, the concrete data of the processes are often only available during the runtime, so that dynamic scheduling must be carried out. The prerequisite for this is, of course, that r_i, Δe_i and d_i at least are known in good time so that the planning step can still take place. For the sake of simplicity, it is therefore assumed below that the time required for real-time scheduling and dispatching itself is negligible (e.g. by planning on an additional processor or special scheduling unit). The result in the case of dynamic scheduling is therefore:

- If

  - EL is a queue with processes that are sorted according to deadlines and are ready (see above),
  - $head(EL)$ is a function that returns the ready process with the shortest deadline (that is, the one that comes first in EL),
  - $insert(EL, i)$ is a function that sorts the process P_i according to its deadline d_i into EL and
  - $dispatch(i)$ function that assigns the processor to the process i.

- The result is Algorithm 13.5.

Algorithm 13.5 EDF for dynamic scheduling.

```
EL = <idle>                                            ▷ rem(idle)=∞
while TRUE do
    p = head(EL)
    dispatch(p)
    arrival= WAIT(ARRIVAL(newp) OR TIME(rem(p)))
    if arrival then
        insert(newp, EL)
    else
        delete(p, EL)
    end if
end while
```

WAIT(ARRIVAL(newp) OR TIME(rest(p))) suspends the scheduling process until either a new process is ready (WAIT returns TRUE) or the time rest(p) has elapsed (WAIT returns FALSE).

EDF is of great importance for real-time calculations. It can be shown that EDF will generate a viable plan, if one exists. With periodic processes, EDF is still able to find a viable plan with any processor utilization

$$u = \sum_{i \in P} \frac{\Delta e_i}{\Delta p_i} \leq 1$$

(with the period Δp_i, s.a.). In other words: As long as the processor utilization u is below 100%, EDF delivers a viable plan, if one exists. The algorithm is therefore optimal in this sense. The problem is that this 100% limit can often not be guaranteed, especially when sporadic processes occur and in multiprocessor systems. If it is exceeded, it is very difficult to predict which processes will miss their deadlines.

13.4.3 Least Laxity

Today's real-time systems are often multiprocessor systems and require adapted scheduling procedures. EDF is not ideal for multiprocessor systems, as the planning does not take into account the point in time at which a process must be scheduled, for example.

Least laxity scheduling supports non-preemptive processes, preemptive processes and multiprocessor systems. As a special feature, preemptive processes can change the processor at any time, and the resulting communication costs are assumed to be either fixed or negligible.

The processes are prioritized according to their *laxity*. This laxity is the difference in time between the ready time and deadline for the process at its execution time.

$$\Delta lax_i = (d_i - r_i) - \Delta e_i$$

For the process to be scheduled in time, the following must apply

$$r_i \leq s_i \leq r_i + \Delta lax_i$$

Here again, a distinction must be made between preemptive and non-preemptive processes, and between static and dynamic scheduling. In the case of dynamic scheduling in preemptive processes, the general solution is considered to be:

- A plan is described by a sequence of tuples $(i, s_{i,j}, \Delta e_{i,j}, m)$ with the processor index m.
- Then the function $llf(P)$ of the set of processes P delivers the process with the smallest laxity and

- the function $active(t, j)$ delivers the process that it active on processor j at the time t.
- It is assumed that P contains as many ready processes as there are processors.
- The procedure is described in Algorithm 13.6.
- The focus here is on the dispatch algorithm. This distributes the new processes from $newSet$ to the processors, as can be seen in Algorithm 13.7.

Algorithm 13.6 Least Laxity.

```
procedure SCHEDULE
  while TRUE do
    llSET = ∅
    readySet = ready(t)
    for j = 1 to m do
      i = llf(readySet)
      llSet = llSet ∪ {i}
      readySet = readySet \ {i}
    end for
    dispatchAll(llSet)
    WAIT(TIME(Δt_slice))
  end while
end procedure
```

Algorithm 13.7 Dispatcher for Least Laxity.

```
procedure DISPATCHALL(llSet)
  activeSet = ∅
  for j = 1 to m do
    activeSet = activeSet ∪ {active(t, j)}
  end for
  newSet = llSet \ activeSet
  for j = 1 to m do
    if active(t, j) ∉ llSet then
      dispatch(get(newSet), j)
    end if
  end for
end procedure
```

This can be seen clearly in the examples in Fig. 13.10.

The algorithm is also easy to display graphically. In addition to the max-flow/min-cut method described in Sect. 13.2.1, there is also a time scale (e.g. as a UML time history diagram, see Sect. 12.2.2):

- At a point in time t, P_i has already passed $\Delta e_i - \Delta rem_i(t)$ of its execution time and $\Delta rem_i(t)$ of its execution time still lies ahead of it. The following therefore applies to the remaining laxity:

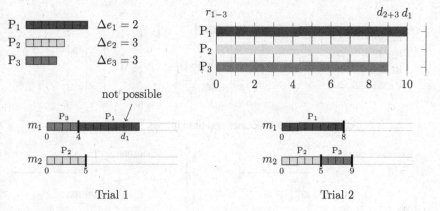

Fig. 13.10 Least laxity scheduling

$$\Delta lax_i(t) = d_i - (t + \Delta rem_i(t))$$

- Let $t_0 = min\{r_1, \ldots, r_n\}$ and m be the number of existing processors.
- On an axis of coordinates, all processes P_i ready at t_0 are entered for the points (x_i, y_i) with

$$x_i : \Delta lax_i(t_0) = d_i - r_i - \Delta e_i$$

$$y_i : \Delta rem_i(t_0) = \Delta e_i$$

- The entries are now changed as follows for each time step Δt_G:
  - A maximum of m ready processes are executed in one step. They move one grid down on the axis of coordinates (corresponds to a reduction of the remaining runtime).
  - All other processes ready for calculation move one grid to the left (corresponds to reduction of the laxity).
  - All new processes that are now ready for calculation are recorded again (see above).

The end occurs when all processes have reached their abscissa (viable plan) or when a process exceeds the ordinate (no viable plan).

As an example, $n = 4$ preemptive processes are to be scheduled for $m = 2$ processors (see Fig. 13.11). The procedure here is to move down those processes that have the smallest laxity. This procedure is complete neither for preemptive nor nonpreemptive processes. This can be seen in the example in Fig. 13.12, in which a viable plan exists (see right side of the image), but it has not been found by the least laxity algorithm (see left side of the image).

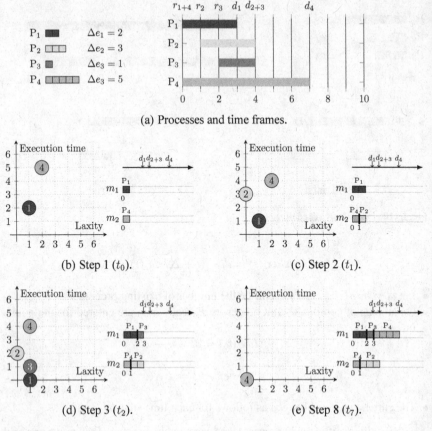

(a) Processes and time frames.

(b) Step 1 (t_0).

(c) Step 2 (t_1).

(d) Step 3 (t_2).

(e) Step 8 (t_7).

Fig. 13.11 Graphical illustration of least laxity scheduling

In a system with m processors, there is a viable plan for the preemptive processes $P = \{P_1, P_2, \ldots, P_n\}$ with the same ready times pd_0 if the following applies for every j with $1 \le j \le k$.

$$\sum_{i \in P} \Delta \mathrm{req}_i(j) \le (pd_j - pd_0)\, m$$

where $\Delta \mathrm{req}_i(j)$ is the minimal required execution time.

In the case of static scheduling, you can assume that the ready times r_i of all processes are the same ($pd_0 = r_i$) and that $P = \{P_1, \ldots, P_n\}$ are already sorted according to deadlines d_i. This results in an algorithm that corresponds to that of EDF for static scheduling of preemptive processes, except that the laxity of the processes must also be adjusted in each step.

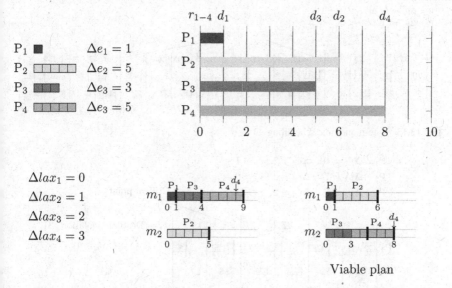

Fig. 13.12 Example for least laxity scheduling: dynamic scheduling, non-preemptive processes

13.4.4 Rate-Monotonic Scheduling

Rate monotonic scheduling (RMS) is suitable for preemptive, periodic processes. A rate is the number of periods in a reference period. There is no explicit plan. Instead, priorities are assigned according to the process rates (implicit plan): prio : $P \Rightarrow Z$. In doing so, processes with shorter periods are preferred over processes with longer periods.

- If $\text{Rate}(P_i) \neq \text{Rate}(P_j)$,
- The priority assignment is carried out anti-proportionally to the period length:
  $\text{rms}(i) < \text{rms}(j) \Leftrightarrow \frac{1}{\Delta p_i} < \frac{1}{\Delta p_j}$,
- with Δp_i: period of P_i and rms: priority for RMS.

The processes are ordered according to the priority determined in this way: $i < j \Leftrightarrow \text{rms}(i) < \text{rms}(j)$. This assignment takes place at certain intervals, the reference periods mentioned above. This is determined, for example, by the longest occurring period of a process.

RMS is not complete, but is very important for real-time systems. The reason for this is its simplicity. In (possible) process changes, only two priority values must be compared: rms(calculating process) and rms(process becoming ready), which is supported by most commercial real-time operating systems.

Of particular importance for RMS is the consideration of critical points in time or critical intervals of the processes, since possible breaches of deadlines can only take place here. The following is defined for this:

$P_1 : \Delta p_1 = 10, \quad \Delta e_1 = 2, \quad r_1 = 1$
$P_2 : \Delta p_2 = 7, \quad \Delta e_2 = 2, \quad r_2 = 0$
$P_3 : \Delta p_3 = 5, \quad \Delta e_3 = 1, \quad r_3 = 1$

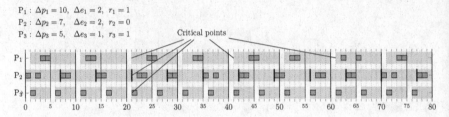

Fig. 13.13 Rate-monotonic scheduling

$P_1 : \Delta p_1 = 10, \quad \Delta e_1 = 2, \quad r_1 = 1$
$P_2 : \Delta p_2 = 7, \quad \Delta e_2 = 2, \quad r_2 = 0$
$P_3 : \Delta p_3 = 5, \quad \Delta e_3 = 2, \quad r_3 = 1$

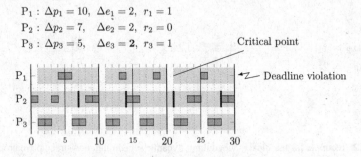

Fig. 13.14 Example 1: Rate-monotonic scheduling with violation of deadlines

- The ready time of P_i in the period k: $r_i^k = r_i + k \cdot \Delta p_i$
- The critical point in time of P_i as the ready time r_i^k in the period k at which the relative completion time $\Delta c_i^k = c_i^k - r_i^k$ is at the maximum, depending on the ready times of all other processes.
- The critical interval of P_i as the interval $[r_i^k, c_i^k]$ at which r_i^k is the critical point in time. See example in Figs. 13.13 and 13.15.
- A sufficient, but not essential condition for P_i is that a critical time r_i^k is always present when the ready times of all higher-priority processes (that is, all processes with shorter periods) fall to r_i^k.

Thus, in the actual scheduling procedure, it only has to be shown that all processes still fit into the respective period in their critical intervals. Although RMS does not consider the importance of the processes but only their period, the priority assignment

$P_1 : \Delta p_1 = 10, \quad \Delta e_1 = 2, \quad r_1 = 1$
$P_2 : \Delta p_2 = 7, \quad \Delta e_2 = 1, \quad r_2 = 0$
$P_3 : \Delta p_3 = 3, \quad \Delta e_3 = 1, \quad r_3 = 0, \ d_3 = 2$

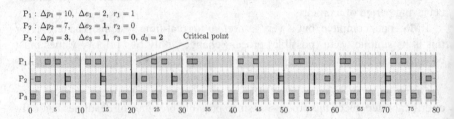

Fig. 13.15 Example 2: Rate-monotonic scheduling

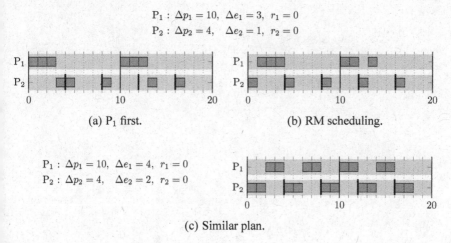

$$P_1 : \Delta p_1 = 10, \quad \Delta e_1 = 3, \quad r_1 = 0$$
$$P_2 : \Delta p_2 = 4, \quad \Delta e_2 = 1, \quad r_2 = 0$$

(a) P_1 first.   (b) RM scheduling.

$$P_1 : \Delta p_1 = 10, \quad \Delta e_1 = 4, \quad r_1 = 0$$
$$P_2 : \Delta p_2 = 4, \quad \Delta e_2 = 2, \quad r_2 = 0$$

(c) Similar plan.

Fig. 13.16 Example 3: Rate-monotonic scheduling

according to RMS is the best method to calculate viable plans (see example in Fig. 13.16). It can generally be shown that, if there is a priority assignment that leads to a viable plan, RMS also leads to a viable plan. An example where this is not possible can be seen in Fig. 13.14. Here P_2 and P_3 are given preference over P_1 at a critical time, so that the latter overruns its deadline.

References

1. Buttazzo GC (2011) Hard real-time computing systems, 3rd edn. Springer, New York
2. Zöbel D (2008) Echtzeitsysteme: Grundlagen der Planung. Springer, Berlin

Appendix A
Physical Units and Quantities

Physical quantities (voltage, time, length etc.) are displayed with a name that represents

- a value and
- a unit of measurement according to the SI system (Système International d'Unités, see Table A.1).

For calculation in quantity equations, the value and unit of measurement are understood as a product: Name = Value · unit of measurement, e.g. voltage $U = 5$·Volts.

To bring units of measurement for a target application into reasonable ranges, units of measurement are given prefixes. A list can be found in Table A.2.

A calculation could then look like this

Table A.1 List of physical quantities

Name	Symbol	Unit	Abbreviation	SI units
Current	I	Ampere	A	SI base unit
Charge	Q	Coulomb	C, As	$1\,C = 1\,As$
Voltage	U	Volt	V	$1\,V = 1\,W/A$
Power	P	Watt	W	$1\,W = 1\,VA$
Resistance	R	Ohm	Ω	$1\,\Omega = 1\,V/A$
Capacity	C	Farad	F	$1\,F = 1\,C/V$
Conductance	G	Siemens	S	$1\,S = 1\,1/\Omega = 1\,A/V$
Energy	W	Joule	J	$1\,J = 1\,Ws$
Temperature	T	Kelvin	K	SI base unit
Time	t	Seconds	s	SI base unit
Length	l	Meter	m	SI base unit

K. Berns et al., *Technical Foundations of Embedded Systems*, Lecture Notes in Electrical Engineering 732, https://doi.org/10.1007/978-3-030-65157-2

Table A.2 Prefixes for units of measurement

Prefix	Meaning	Factor
T	Tera	10^{12}
G	Giga	10^{9}
M	Mega	10^{6}
k	Kilo	10^{3}
d	Deci	10^{-1}
c	Centi	10^{-2}
m	Milli	10^{-3}
μ	Micro	10^{-6}
n	Nano	10^{-9}
p	Pico	10^{-12}
f	Femto	10^{-15}

$$
\begin{aligned}
U &= 1\,\mu\text{A} \cdot 50\,\text{m}\Omega \\
&= 1 \cdot 10^{-6} \cdot 50 \cdot [10^{-3}]\,\text{A} \cdot \Omega \\
&= 50 \cdot 10^{-9}\,\text{V} \\
&= 50\,\text{nV}
\end{aligned}
$$

Appendix B
Resistance Values, Circuit Symbols and Diode Types

B.1 Specific Resistance

The resistance of a conductor at room temperature depends on the material, the length l of the conductor and its cross-sectional area q. This material dependence is called *specific resistance* ρ. The resistance is then calculated as

$$R = \rho \cdot \frac{l}{q}$$

The specific resistances of different materials are given in Table B.1.

Table B.1 Specific resistances of different materials at 20 °C

Material	Spec. resistance ρ [$\frac{\Omega \cdot mm^2}{m}$]
Silver	0.016
Copper	0.018
Gold	0.022
Aluminum	0.028
Zinc	0.06
Brass	0.07
Iron	0.1
Platinum	0.106
Tin	0.11
Lead	0.208
Coal	66.667

© The Author(s), under exclusive license to Springer Nature Switzerland AG 2021
K. Berns et al., *Technical Foundations of Embedded Systems*, Lecture Notes in Electrical
Engineering 732, https://doi.org/10.1007/978-3-030-65157-2

Table B.2 Color coding of resistors with 4 rings

Color	1. Ring	2. Ring	3. Ring	4. Ring
Silver	–	–	$\times 10^{-2}$	$\pm 10\%$
Gold	–	–	$\times 10^{-1}$	$\pm 5\%$
Black	–	0	$\times 10^{0}$	–
Brown	1	1	$\times 10^{1}$	$\pm 1\%$
Red	2	2	$\times 10^{2}$	$\pm 2\%$
Orange	3	3	$\times 10^{3}$	–
Yellow	4	4	$\times 10^{4}$	–
Green	5	5	$\times 10^{5}$	$\pm 0.5\%$
Blue	6	6	$\times 10^{6}$	$\pm 0.25\%$
Purple	7	7	$\times 10^{7}$	$\pm 0.1\%$
Grey	8	8	$\times 10^{8}$	$\pm 0.05\%$
White	9	9	$\times 10^{9}$	–

B.2 Color Coding of Resistors

Resistors are color-coded as electrical components. There are color codes with four to six rings. With four rings, the 1st and 2nd ring indicate the numerical value, the 3rd the multiplier and the 4th ring the tolerance class. The exact color coding can be found in Table B.2. For example, the coding *yellow-purple-red-gold* stands for $47 \cdot 10^{2}\ \Omega$ with a tolerance of $\pm 5\%$.

B.3 Circuit Symbols

Although a complete overview of all circuit symbols cannot be given here, at least the most important ones are shown in Table B.3.

The circuit symbols presented here comply with the current DIN standard, but there are still many outdated or unofficial circuit symbols in circulation. In this book, for example, a direct voltage source is often represented by a circle with a "=" sign in the middle, so that it can be distinguished directly from an AC voltage source. The standard symbol for inductances (on the left in Table B.3) is rarely found in textbooks, while the outdated one (on the right of the figure) still appears frequently. This inconsistency is due to historical development.

Table B.3 Circuit symbols

Component	Circuit symbol	Component	Circuit symbol
Resistor (or resistance), general		NPN transistor	
Resistor (or resistance), changeable (Potentiometer)		PNP transistor	
Capacitor		Voltage source, ideal	
Coil or inductivity		Current source, ideal/constant	
Signal lamp off/on		AC voltage	
Motors		DC voltage	
Voltmeter		Jump signal generator	
Diode		Battery	
LED		Ground	
Photo diode		Operational amplifier	

B.4 Diode Types

When a diode is operated in the forward range, the diode voltages U_F result for the typical currents I_F. For reverse operation, however, the reverse currents I_R for the applied diode voltages U_R (negative value) are used. In addition, it must be noted that a voltage in reverse direction that is greater than U_{Rmax} will destroy normal diodes. In the passband the current I_{Fmax} is the limit, as with a larger current the diode gets

Table B.4 Properties of a selection of diode types

	1N914	AA139	1N4005	SSik2380	SSiP11
Application	Switch	Universal	Rectifier	Power	Power
Material	Si	Ge	Si	Si	Si
Parameters					
Voltage U_{F_1}	0.6 V	0.21 V	0.8 V	0.85 V	0.82 V
at current I_{F_1}	1 mA	1 mA	10 mA	10 A	100 A
Voltage U_{F_2}	1.0 V	0.5 V	1.0 V	1.02 V	1.15 V
at current I_{F_2}	100 mA	100 mA	1 A	100 A	1500 A
Blocking current I_R	25 nA	1 μA	10 μA	15 mA	60 mA
at U_R	20 V	10 V	600 V	1400 V	2200 V
at T	25 °C	25 °C	25 °C	180 °C	180 °C
Limits					
$I_{F_{max}}$	200 mA	200 mA	1.0 A	160 A	1500 A
$U_{R_{max}}$	75 V	20 V	600 V	1400 V	2200 V

too hot and is destroyed by it. Exemplary values for some commercially available diode types are given in Table B.4.

Appendix C
Fourier and Laplace Tables

C.1 Examples of Fourier Series

In the following Table C.1 you will find some examples of Fourier series with their properties. The series are not always specified completely and their continuation is indicated by (...).[1]

C.2 Fourier Transformation Table

Table C.2 contains the Fourier transforms and inverse transforms for the most important elementary functions. By decomposing more complex functions into a sum of these elementary functions (e.g. by means of the residual theorem or a partial fractional decomposition), the respective transform can be determined.

C.3 Laplace Transformation Table

Table C.3 contains the Fourier transforms and inverse transforms for the most important elementary functions. By decomposing more complex functions into a sum of these elementary functions (e.g. by means of the residual theorem or a partial fractional decomposition), the respective transform can be determined.

---

[1] Source: [1].

© The Author(s), under exclusive license to Springer Nature Switzerland AG 2021   369
K. Berns et al., *Technical Foundations of Embedded Systems*, Lecture Notes in Electrical
Engineering 732, https://doi.org/10.1007/978-3-030-65157-2

Table C.1 Examples of the Fourier Series

$\omega = \dfrac{2\pi}{T}$

1		Antisymmetric rectangular function, duty cycle 0.5, no direct component
	$f(t) = A \cdot \frac{4}{\pi} \left(\sin \omega t + \frac{1}{3} \sin 3\omega t + \frac{1}{5} \sin 5\omega t + \cdots \right)$	
2		Symmetric rectangular function, duty cycle 0.5, no direct component
	$f(t) = A \cdot \frac{4}{\pi} \left(\cos \omega t - \frac{1}{3} \cos 3\omega t + \frac{1}{5} \cos 5\omega t - \cdots \right)$	
3		Square pulse, duty cycle τ/T
	$f(t) = A \cdot \frac{\tau}{T} + A \cdot \frac{2}{\pi} \cdot \left(\sin \pi \frac{\tau}{T} \cdot \cos \omega t + \frac{1}{2} \sin \pi \frac{2\tau}{T} \cdot \cos 2\omega t + \cdots \right)$	
4		Bipolar Square pulse, half cycle symmetry, proxy $\varphi = 2\pi\tau/T$
	$f(t) = A \cdot \frac{4}{\pi} \left(\frac{\cos \varphi}{1} \sin \omega t + \frac{\cos 3\varphi}{3} \sin 3\omega t + \frac{\cos 5\varphi}{5} \sin 5\omega t + \cdots \right)$	
5		Trapezoid oscillation, rise time = fall time = τ, proxy $a = 2\pi\tau/T$
	$f(t) = \frac{A}{a} \cdot \frac{4}{\pi} \left(\frac{\sin a}{1^2} \sin \omega t + \frac{\sin 3a}{3^2} \sin 3\omega t + \frac{\sin 5a}{5^2} \sin 5\omega t + \cdots \right)$	
6		Antisymmetric triangular oscillation, half cycle symmetry, no direct component
	$f(t) = A \cdot \frac{8}{\pi^2} \left(\sin \omega t - \frac{1}{3^2} \sin 3\omega t + \frac{1}{5^2} \sin 5\omega t - \cdots \right)$	
7		Symmetric triangular oscillation, half cycle symmetry, no direct component
	$f(t) = A \cdot \frac{8}{\pi^2} \left(\cos \omega t + \frac{1}{3^2} \cos 3\omega t + \frac{1}{5^2} \cos 5\omega t + \cdots \right)$	

(continued)

Table C.1 (continued)

$$\omega = \frac{2\pi}{T}$$

8	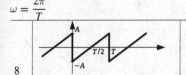	Saw tooth oscillation, no direct component, anti-symmetry
	$f(t) = A \cdot \frac{2}{\pi} \left(\sin \omega t + \frac{1}{2} \sin 2\omega t + \frac{1}{3} \sin 3\omega t + \cdots \right)$	
9	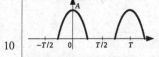	Cosine oscillation after after both sided rectification, whole cycle symmetry, T: period of the mains frequency
	$f(t) = A \cdot \frac{2}{\pi} + A \cdot \frac{4}{\pi} \cdot \left(\frac{1}{1\cdot3} \cos 2\omega t - \frac{1}{3\cdot5} \cos 4\omega t + \frac{1}{5\cdot7} \cos 6\omega t - \cdots \right)$	
10	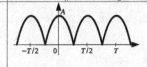	Cosine oscillation after one-sided rectification
	$f(t) = A \cdot \frac{1}{\pi} + A \cdot \frac{2}{\pi} \cdot \left(\frac{\pi}{4} \cos \omega t + \frac{1}{1\cdot3} \cos 2\omega t - \frac{1}{3\cdot5} \cos 4\omega t + \frac{1}{5\cdot7} \cos 6\omega t - \cdots \right)$	
11	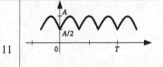	Rectified three-phase current, T: period of the mains frequency
	$f(t) = A \cdot \frac{3\sqrt{3}}{\pi} \cdot \left(\frac{1}{2} - \frac{1}{2\cdot4} \cos 3\omega t - \frac{1}{5\cdot7} \cos 6\omega t - \frac{1}{8\cdot10} \cos 9\omega t - \cdots \right)$	

Table C.2 Fourier transformation table (excerpt)

$f(t)$	$F(\omega)$
$\delta(t)$	1
1	$2\pi \cdot \delta(\omega)$
$\sigma(t)$	$\frac{1}{j\omega} + \pi\delta(\omega)$
$sgn(t)$	$\frac{2}{j\omega}$
$e^{j\omega_0 t}$	$2\pi \cdot \delta(\omega - \omega_0)$
$sin(\omega_0 t)$	$j\pi \cdot [\delta(\omega + \omega_0) - \delta(\omega - \omega_0)]$
$cos(\omega_0 t)$	$\pi \cdot [\delta(\omega + \omega_0) + \delta(\omega - \omega_0)]$
$rect\left(\frac{t}{2T}\right) = \begin{cases} 1, & \text{for } |t| \leq T \\ 0, & \text{for } |t| > T \end{cases}$	$2T \cdot \frac{\sin(\omega T)}{\omega T} = 2T \cdot si(\omega T)$
$e^{-at} \cdot \sigma(t)$	$\frac{1}{a + j\omega}$
$e^{-a|t|} \cdot cos(bt)$	$\frac{2a \cdot (a^2 + b^2 + \omega^2)}{(a^2 + b^2 + \omega^2)^2 + 4a^2\omega^2}$
$e^{-a|t|} \cdot sin(bt)$	$\frac{-j \cdot 4ab\omega}{(a^2 + b^2 - \omega^2)^2 + 4a^2\omega^2}$

Table C.3 Laplace transformation table

Nr.	$f(t)$	$F(s)$
1	$\delta(t)$	1
2	$\sigma(t)$	$\frac{1}{s}$
3	t	$\frac{1}{s^2}$
4	$\frac{t^{n-1}}{(n-1)!}$	$\frac{1}{s^n}$
5	e^{-at}	$\frac{1}{s+a}$
6	$t \cdot e^{-at}$	$\frac{1}{(s+a)^2}$
7	$t^2 \cdot e^{-at}$	$\frac{2}{(s+a)^3}$
8	$1 - e^{-at}$	$\frac{a}{s \cdot (s+a)}$
9	$\sin(\omega t)$	$\frac{\omega}{s^2+\omega^2}$
10	$\cos(\omega t)$	$\frac{s}{s^2+\omega^2}$
11	$e^{-at}\sin(\omega t)$	$\frac{\omega}{(s+a)^2+\omega^2}$
12	$e^{-at}\cos(\omega t)$	$\frac{s+a}{(s+a)^2+\omega^2}$

Reference

1. Kories R, Schmidt-Walter H (2011) Electrical engineering: a pocket reference. Springer Science & Business Media, Frankfurt a. M

Index

© The Author(s), under exclusive license to Springer Nature Switzerland AG 2021
K. Berns et al., *Technical Foundations of Embedded Systems*, Lecture Notes in Electrical Engineering 732, https://doi.org/10.1007/978-3-030-65157-2

Printed in the United States
by Baker & Taylor Publisher Services